AIR CONDITIONING AND REFRIGERATION REPAIR

SECOND EDITION

AIR CONDITIONING AND REFRIGERATION REPAIR
SECOND EDITION

ROGER A. FISCHER AND KEN CHERNOFF

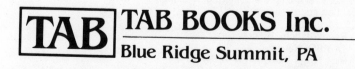

TAB BOOKS Inc.
Blue Ridge Summit, PA

SECOND EDITION
SECOND PRINTING

Copyright © 1988 by TAB BOOKS Inc.
Printed in the United States of America

Library of Congress Cataloging in Publication Data

Fischer, Roger A.
     Air conditioning and refrigeration repair / by Roger A. Fischer
and Ken Chernoff. — 2nd ed.
     p. cm.
     Rev. ed. of: Successful air conditioning & refrigeration repair.
1st ed. c1982.
     Includes index.
     ISBN 0-8306-9581-8   ISBN 0-8306-2881-9 (pbk.)
     1. Refrigeration and refrigerating machinery—Maintenance and
repair. 2. Air conditioning—Equipment and supplies—Maintenance
and repair. I. Chernoff, Ken. II. Fischer, Roger A. Successful
air conditioning & refrigeration repair. III. Title.
TP492.7.F57 1988
621.5'6—dc 19   88-15988
                                                           CIP

TAB BOOKS Inc. offers software for sale. For information and a catalog, please contact TAB Software Department, Blue Ridge Summit, PA 17294-0850.

Questions regarding the content of this book
should be addressed to:

     Reader Inquiry Branch
     TAB BOOKS Inc.
     Blue Ridge Summit, PA 17294-0214

Contents

Tool . . . Inspection Mirror . . . Service-Valve Wrench . . .
Crimping Tool

Preface

This is a practical book, written by a man who presently works in the field and has instructed in the field as well as in the classroom. This book is a reference manual that can be carried in your tool box. Much of the information was gained through work experience and does not appear in any other text or reference manual.

Students bring many different books on class field problems with them. These texts and reference books usually contain many pages with a lot of words that can not be related to the field. These books are merely theory presented in classroom lectures. Students want a "hands-on" reference guide that can easily be carried. This book fills that need.

Handy homeowners will find this book a helpful guide to making simple repairs and understanding how air conditioning works. Performing simple preventive maintenance procedures can save you—the homeowner—money, as well as spare you many unnecessary breakdowns.

Acknowledgments

A very special thanks to R.M. Ward & Sons, Inc., of Oldsmar Florida, whose cooperation shall never be forgotten.

Thanks also to H.B. Adams Inc., of Clearwater, Florida.

Introduction

Contributions to this book were written by dedicated men who worked most of their lives in the refrigeration and air conditioning fields. This edition of *Air Conditioning and Refrigeration Repair* is targeted to the frustrated service technicians and homeowners who deal with malfunctioning equipment. We have seen people, during the heat of summer, enduring torturous conditions waiting for a service technician to arrive. In many cases, a better understanding of this equipment could allow its repair to be made by the homeowner.

The text covers residential equipment only, including window units, automobile units, and refrigerators. Some light commercial equipment is discussed. For the first time, text has been prepared with both the homeowner and professional in mind. Its method of presentation does not demean the intellect of either; yet both will understand thoroughly how the system works.

This book explains in detail all the preventive maintenance that can be performed by homeowners to add years to the lives of their systems. Service technicians can use this guide to establish preventive maintenance programs for their customers.

This book can be carried in the technician's tool box or placed in the home library by the homeowner. Should the homeowner save the cost of one service call and all the discomfort connected with the malfunction, this book has served its purpose.

CHAPTER 1

Basic Electricity for Refrigeration

I want to cover in a practical way the failures and troubles that occur in the air conditioning field. If you analyze air conditioning failures, you find that 80 percent of them are electrical failures. The remaining 20 percent of the problems are in the closed refrigerant system. With this in mind, the book is focused on 80 percent electrical malfunctions and 20 percent on closed system failure. If you look at the books on library shelves concerning air conditioning and refrigeration, much of the texts are not relevant to the field. They might have one or two chapters on electrical failures. I have written this book so the student or professional will be able to go into the field with a few tools and diagnose problems.

With this in mind, I begin with basic electricity. High-voltage transmission wires are carried within insulators on top of cross-members of power poles. Voltages are quite high; they can be 4160 volts, 3600 volts, 2400 volts, or higher. The purpose of using high voltage is to reduce wire resistance line losses.

Figure 1-1 shows distribution from network to step-down transformer with the primary attached to the high-voltage top cross bar conductors. The secondary of this transformer has a center-tap neutral and two line wires. Line one and line two are connected to the lower conductors on the bottom cross bars of the power poles. In the schematic, the secondary wires are leaving the transformer and are wired to an electrical watt-hour meter that records energy consumed.

1

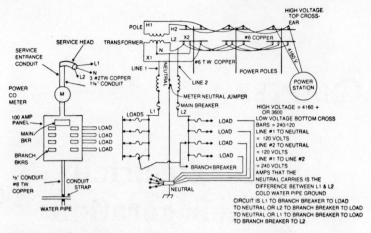

Fig. 1-1. Electricity distribution from the network to the step-down transformer.

From the bottom of the meter, the conductors are wired to a main circuit breaker or fuses. Another conductor is brought through the meter with the use of a jumper wire. It is connected to a neutral bar in the entrance panel. The electric outlet side of the main circuit breaker is called the *load side*. The inlet side of the circuit breaker that is wired to the meter is the *line side*. The line side to a fuse or circuit breaker is always energized with electrical power as long as it is being supplied by the power company. The load side of the main circuit breaker is connected to two conductors called the bus. The bus is the main supply to the branch circuit breakers line side. Figures 1-2 through 1-6 show actual photographs of the components mentioned in Fig. 1-1.

Figure 1-7 shows a typical circuit breaker entrance panel with the cover removed. You can see the bus bars exposed where the circuit breakers were removed. The voltage/current flows from the bus through the circuit breaker of the branch circuit to the load (electrical consuming device-appliance-lighting). It then returns to the panel by way of the neutral, if the voltage is 120 volts. If the voltage is 240 volts, single phase, it will leave a branch circuit breaker on line one, flow to the load, and return to the panel on the load side of the same breaker on line two bus. The voltage from line one to neutral is 120 volts. The voltage from line two to neutral is 120 volts. The voltage from line one to line two is 240 volts. The current that the neutral carries in a three wire circuit is the difference between the currents in line one and line two.

2

Fig. 1-2. Generating plant where electricity is produced and transmitted through the network.

The cold-water-pipe ground is wired to the neutral bar of the entrance panel. As long as the neutral is wired to the cold-water-pipe ground, the voltage of the neutral will be zero volts in respect to ground. The ground wire that connects to the load also connects to the cold-water ground at the panel.

The purpose of the service head (Fig. 1-5) is to keep water from entering the service-entrance conduit, meter, and main panel. The purpose of the service-entrance conduit is to protect the service-entrance conductors from physical damage and, in the case of fire, to protect against electrocution. Conduit is also used as a ground conductor. In many cases, branch circuit wiring leaves the main panel enclosed in conduit. Some types of conduit are rigid (hard wall), flexible, or thin wall. Romex is a wiring system using a conductor made of fiber or heavy plastic covering. In some areas, knob-and-tube type of wiring may still be seen (most areas have outlawed it).

The step-down transformer shown in Fig. 1-4 is normally mounted on the power pole. It is oil-filled and air-cooled. This type of transformer reduces voltages to the amount needed for customer purchase. Most homes today are a three-wire system. You can see them entering the structure. This type of arrangement allows the use of both 120- and 240-volt, single phase ac appliances. On some older homes built over 45 years ago, you might find two wires; this can only supply 120 volts ac, single-phase (only one line and the

Fig. 1-3. Substation where electricity is passed through step-down transformers before being sent through the local network.

neutral was used). Referring to the schematic in Fig. 1-1, you can see that the wires from the pole feed the meter first. The meter is in series with the main breaker which is in series with all the branch circuit breakers of the entrance panel. The branch circuit breaker is the last over-current device that feeds an electrical load. In the

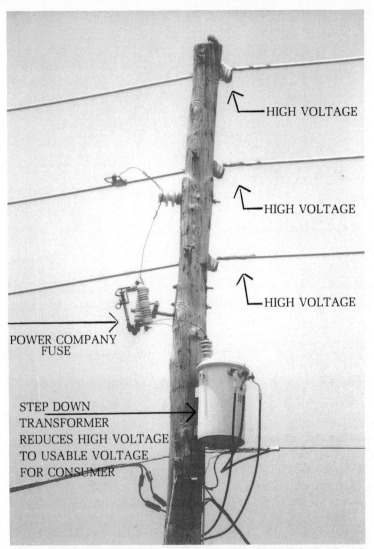

HIGH VOLTAGE

HIGH VOLTAGE

HIGH VOLTAGE

POWER COMPANY FUSE

STEP DOWN TRANSFORMER REDUCES HIGH VOLTAGE TO USABLE VOLTAGE FOR CONSUMER

Fig. 1-4. Typical step-down transformer found in the local neighborhood where the electrical voltage is dropped to the amount used by the consumer.

case of 120 volts, the electrical flow is to the load and returns to the panel to the neutral. With 240 volts, electricity flows to the load through line one and returns to the branch breaker through line two.

Most changes or modifications in wiring deal with branch circuit wiring or feeder wiring. A *feeder wire* is a wire that connects be-

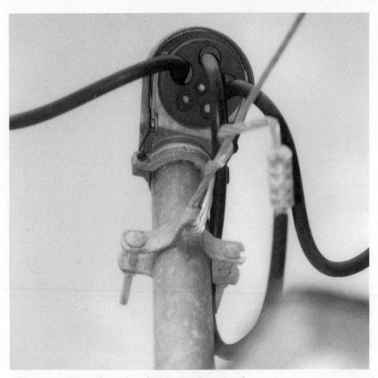

Fig. 1-5. Weatherproof service head where power company brings their transmission lines to the structure being supplied.

tween a branch breaker and a main panel breaker. You will find that branch circuit wiring can be many different sizes of wire as well as different amounts of conductors encased in the conduit. The size of the wire is determined by the maximum amperage consumption of the electrical load that the wire will supply.

For example, if you have a load (machine, appliance, etc.) that needs a current of 20 amps, according to the electrical code, you will need a #12 TW wire (the number of the wire refers to the American Wire Gauge). The lower the number, the larger the diameter becomes. The larger the diameter, the more current (load) the wire can carry safely. Table 1-1 shows an ampacity chart for copper conductors.

Due to the fact aluminum does not conduct as well as copper, aluminum wiring of the same gauge is rated to carry less current. Table 1-2 shows ampacity for aluminum conductors, and the difference between the two conductors is easily seen.

Fig. 1-6. Electric meter supplied by the power company, measures the amount of power being consumed within the structure.

The diameter of conduit is dependent upon the region in which you are working. Consult the NEC (National Electrical Code) and the local electrical codes of your area. The amount of wires placed in a conduit is called the *conductor fill*. Some areas have a specified number of conductors allowed, depending upon the wire size. Other places allow as many conductors into a conduit as you can push through. Table 1-3 is a sample conductor fill chart that is used in some areas.

The amount of current a wire can carry is dependent upon the type of metal the wire is made of, diameter of the wire, type of insulation covering the wire, and whether the wire is suspended in free air or inside an electrical conduit. High-voltage wiring normally has a small current. For this reason, the high voltage transmission lines on the power poles have small wiring.

Fig. 1-7. Typical circuit breaker entrance panel with cover removed.

You are charged for electricity by the amount you use. The unit of measure is watt-hours. Example: A one horsepower motor which is equal to 800 watts costs the same to operate at 120 volts as it does if you operate it at 240 volts. The 800 watt-hours is the same regardless of the voltage.

8

Table 1-1. Ampacity of Copper Conductors.

SIZE	NAME* TW	NAME RH, RHW, THW, THHN.	NAME AVA	NAME A
14	15	15	30	30
12	20	20	35	40
10	30	30	45	55
8	40	45	60	75
6	55	65	80	95
4	70	85	105	120
2	95	115	135	165
1	110	130	160	190
0	125	150	190	225
00	145	175	215	250
000	165	200	245	285
0000	195	230	275	340
250	215	255	315	----
300	240	285	345	----
350	260	310	390	----
400	280	335	420	----
500	320	380	470	----
600	355	420	525	----
700	385	460	560	----
750	400	475	580	----
800	410	490	600	----
900	435	520	----	----
1000	455	545	680	----
1250	495	590	----	----
1500	520	625	785	----
1750	545	650	----	----
2000	560	665	840	----

*Name codes are given in Table 1-5

Example:

$$\text{Watts} = \text{volts} \times \text{amps}$$

1 hp

120 volts

$$800 = 120 \times \text{amps}$$

$$120 = 120$$

$$\frac{800}{120} = \text{amps}$$

$$6\frac{2}{3} = \text{amps}$$

Table 1-2. Ampacity of Aluminum Conductors.

SIZE	NAME* TW	NAME RH, RHW THW, THW-N	NAME AVA	NAME A
12	15	15	25	30
10	25	25	35	45
8	30	40	45	55
6	40	50	60	75
4	55	65	80	95
2	75	90	105	130
1	85	100	125	150
0	100	120	150	180
00	115	135	170	200
000	130	155	195	225
0000	155	180	215	270
250	170	205	250	----
300	190	230	275	----
350	210	250	310	----
400	225	270	335	----
500	260	310	380	----
600	285	340	425	----
700	310	375	455	----
750	320	385	470	----
800	330	395	485	----
900	355	425	----	----
1000	375	445	560	----
1250	405	485	----	----
1500	435	520	650	----
1750	455	545	----	----
2000	470	560	705	----

*Name codes are given in Table 1-5

Example:

$$800 = 240 \times amps$$

$$\begin{matrix} 1 \text{ hp} \\ 240 \text{ volts} \end{matrix} \qquad \frac{800}{240} = \frac{240 \times amps}{240 \times amps}$$

$$amps = 3\frac{1}{3}$$

$$120 \text{ volts} \times 6\frac{2}{3} \text{ amps} = 240 \text{ volts} \times 3\frac{1}{3}$$
$$800 \text{ watts} = 800 \text{ watts}$$

You pay for watt-hours (watts × time in hours). Some salesmen will attempt to sell a higher voltage machine of the same horsepower with the understanding that due to the lower current draw, the running cost will be less. This whole line of reasoning is false.

Table 1-3. Sample Conductor Fill Chart for EMT Used in Certain Areas Where Electrical Code is Still in Force.

BASED UPON 40% CONDUCTOR FILL WITHIN THE INSIDE DIAMETER OF CONDUIT

WIRE SIZE	WIRES IN WIRE BODY	WIRE DIAMETER INCHES	WIRE NAME	WIRE SIZE	½	¾	1	1¼	1½	2	2½	3	3½	4
18	1	.0403	TW											
16	1	.0508	THW											
14	1	.0641	THHN	14	9	15	25	44						
12	1	.0808		12	7	12	19	35						
10	1	.1019		10	5	9	15	26						
8	7	.1285		8	3	5	8	14						
6	7	.0612		6	1	2	4	7						
4	7	.0772		4	1	1	3	5						
2	7	.0974		2	1	1	2	4	4					
1	19	.0664		1		1	1	3	3					
0	19	.0745		0		1	1	2	3	4				
00	19	.0837		00		1	1	1	2	3				
000	19	.0940		000			1	1	1	2	4			
0000	19	.1055		0000			1	1	1	2	3			
250	37	.0822		250				1	1	1	3			
300	37	.0900		300				1	1	1	2			
350	37	.0973		350				1	1	1	1	4		
400	37	.1040		400				1	1	1	1	4		
500	37	.1162		500				1	1	1	1	3	4	
600	61	.0992		600								3	4	
700	61	.1071		700								2	3	4
750	61	.1109		750								2	3	4
800	61	.1145												
900	61	.1215												
1000	61	.1280												
1250	91	.1172												
1500	91	.1284												
1750	127	.1174												
2000	127	.1255												

WIRE PHYSICAL INDENTIFICATION

BY USING A MICROMETER ON EACH WIRE STRAND IN THE WIRE BODY, YOU WILL BE ABLE TO IDENTIFY WIRE SIZE. YOU MUST KNOW WIRE SIZE TO DETERMINE IF THE CIRCUIT IS OVERLOADED.

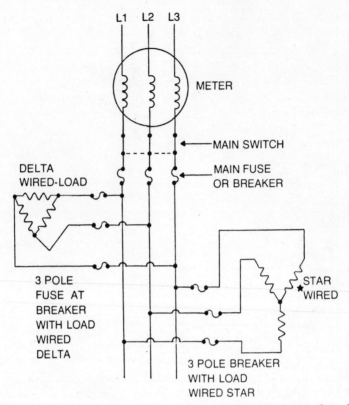

Fig. 1-8. Schematic of a three-phase distribution panel and meter.

Figure 1-8 is a schematic of a three-phase power distribution panel and meter. Lines one, two, and three all have voltage when energized. Voltage can be 208, 240, 480, or higher. Note the *delta-wired* load on the left of the drawing and the *star-wired* load on the right of the drawing. All three-phase loads are wired either delta or star.

Three-phase distribution uses three lines in which the peak voltage in each conductor is reached successively one-third of a cycle apart, 120 degrees of generator rotation. In effect there are three separate sine waves of potential (voltage) and three waves of current per cycle.

The current and voltage sine waves of a resistance circuit will be in step with each other. Normally, they have different wave heights, but the voltage wave intersects the amperage wave at 0, 180, and 360 degrees of phase time. When a coil (inductance) is in the circuit, the voltage sine wave will lead the current sine wave, or reach its peak value earlier in time.

With a capacitor in the circuit, the current will peak before the voltage, or the current wave will lead the voltage sine wave. An inductive circuit may have a capacitor placed in parallel to increase the power factor. With proper balance, the power factor can be corrected above nine-tenths; this process can also be reversed. A low power factor means the efficiency of the circuit is low and you end up paying for power that does no economic work.

ELECTRICITY

Electricity is a form of energy that is a result of electrons in motion. The three important properties of electricity are voltage, current, and resistance. *Voltage* is electromotive force that pushes a flow of electrons, known as *current*, through a conductor or medium that has a *resistance* to slow the electrons that are being pushed.

OHM'S LAW

Ohm's law states that an electromotive force of one volt is required to push a current of one ampere through a resistance of one ohm. Voltage is a measure of electromotive force, potential, or electric pressure. Current is the volume of the flow of electrons being pushed along a conductor. Resistance is the ability of a substance or conductor to impede the flow of electrons being pushed by voltage. Power is the capacity of electricity to do work in a given time period and is expressed in watts. Letter symbols are as follows:

E or V designates voltage, electromotive force, electrical pressure
I   designates current
R   designates resistance
W   designates power, in watts

The triangle shown in Fig. 1-9 illustrates Ohm's law. Use this triangle to find the unknown factor when two factors are known.

Step 1—Cover the electrical value you want to find. The relationship of the formula becomes apparent. Example: If you want the value for E (voltage), cover E on the triangle.

Step 2—You can see I × R is visible. That is the formula. If you want to find I (current), cover I in the triangle. The formula relationship now would be $\dfrac{E}{R}$ to find I. To solve for R (resistance),

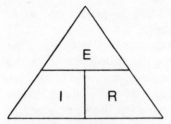

Fig. 1-9. Ohm's law triangle.

cover R in the triangle. The formula relationship now becomes $\frac{E}{I}$ to find R. You will need these formulas often.

Circuitry

The two basic circuits in electricity are the *series* circuit and the *parallel* circuit. At the top of Fig. 1-10 a basic series circuit is shown. You can see there is only one path for the electricity to flow through from L1 to L2. If a device is placed into this type of circuit, the electrical energy will flow through it. In the bottom section of Fig. 1-10 an example of a basic parallel circuit is shown. You can see in this circuit, there are three different paths through which the electrical energy can flow at the same time. Both of these circuits are resistive as shown by the resistors labeled A, B, and C.

To find total resistance in the circuit use the following procedure. In the series circuit, just add the values of each resistor, and the total amount found is the total amount of resistance in the circuit. In the parallel circuit the relationship is different. The total resistance of a parallel circuit is usually smaller than the value of the smallest resistor in the circuit. In Fig. 1-11 you will find total resistance for the circuit to be smaller than the smallest resistor.

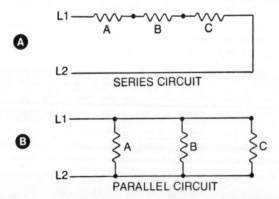

Fig. 1-10. (top) Series circuit; (bottom) Parallel circuit.

14

$$\text{MILLI} = \frac{1}{1,000} = .001 \text{ TIMES UNITS OF MEASURE.}$$
EXAMPLE: MILLIVOLTS, MILLIAMPS, MILLIWATTS.

$$\text{MICRO} = \frac{1}{1,000,000} = .000001 \text{ TIMES UNIT OF MEASURE.}$$
EXAMPLE: MICROVOLTS, MICROAMPS, MICROWATTS.

KILO = 1,000 TIMES THE UNIT OF MEASURE.
EXAMPLE: KILOVOLTS, KILOWATTS

MEGA = 1,000,000 TIMES THE UNIT OF MEASURE.
EXAMPLE: MEGAWATTS OR MEGOHMS.

RESISTANCE SERIES FORMULA: $R_t = R_1 + R_2 + R_3$

EXAMPLE: R_1 = 10 OHMS $R_t = 10 + 5 + 1 = 16$ OHMS
R_2 = 5 OHMS
R_3 = 1 OHM

(R_t STANDS FOR RESISTANCE TOTAL).

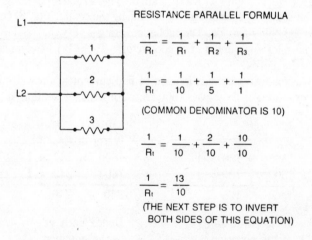

RESISTANCE PARALLEL FORMULA

$$\frac{1}{R_t} = \frac{1}{R_1} + \frac{1}{R_2} + \frac{1}{R_3}$$

$$\frac{1}{R_t} = \frac{1}{10} + \frac{1}{5} + \frac{1}{1}$$

(COMMON DENOMINATOR IS 10)

$$\frac{1}{R_t} = \frac{1}{10} + \frac{2}{10} + \frac{10}{10}$$

$$\frac{1}{R_t} = \frac{13}{10}$$

(THE NEXT STEP IS TO INVERT
BOTH SIDES OF THIS EQUATION)

Fig. 1-11. Finding resistance totals in series and parallel circuits.

Conductors

All general building wiring is rated at 600 volts ac. The current rating depends on the wire size (diameter of the wire) or the wire gauge, type of material (copper, aluminum), and type of insulation. Wiring is sized to handle the current consumption of the load. See Tables 1-1 and 1-2.

Conduit is a metal enclosure that protects the wiring and allows for a safe grounding. It is commonly called EMT (extruded metallic tubing). A nonmetallic conduit has been developed and accepted by the building industry which is accepted in many local codes, but of course it lacks the protection and grounding ability of the EMT. This plastic conduit is called PVC. In certain applications. Romex (a fibrous-covered wire) and cords may be used without conduit.

Grounding of PVC conduit is accomplished with the addition of a green-colored conductor that is used for grounding only. In the EMT system, grounding is accomplished by the metal of the conduit that is bonded to the grounding conduit to the cold water pipe.

Many small compressor/condenser sections are wired with a cord instead of conduit. Cords are rated at 250 volts or 600 volts. Check NEC and local codes for maximum allowable length of connecting cords. In some areas, the limit is 10 feet. Cords are coded to identify construction and application. As an example, cords may be coded SO, SJO, or SJ. SO cord has a heavy-duty, thick rubber covering with a jute filler surrounding the conductors. SJO cord is rated for medium duty with a medium thickness rubber covering and a combination of paper and jute surrounding the conductors. SJ cords are for light duty service. The rubber covering is very thin, and there is paper filler surrounding the conductors. Table 1-4 has ampacity ratings for cords.

The following formula is used to find current when watts and volts are known. Example: 1600 watt compressor section at,

120 volts

$$Watts = E \times I$$
$$1600 = 120 \times I$$
$$\frac{1600}{120} = \frac{120 \times I}{120}$$

Divide both sides
of the equation
by 120.
$I = 13.3$ amps

240 volts

$$Watts = E \times I$$
$$1600 = 240 \times I$$
$$\frac{1600}{240} = \frac{240 \times I}{240}$$

Divide both sides
of the equation
by 240.
$I = 6.7$ amps

Table 1-4. Wire Cords.

SIZE	NAME* S, SO, SJ, SJO, THERMO- PLASTIC	WATTS VOLTAGE 120V/240V
18	7 AMPS	840/1180
16	10	1200/2400
14	15	1800/3600
12	20	2400/4800
10	25	3000/6000
8	35	4200/8400
6	45	5400/10800
4	60	7200/14400
2	80	9600/19200

CORD RATED AT 250 or 600
VOLTS ac SEE CORD
LABEL. SMALL COMPRESSOR
CONDENSER UNITS USE CORD AND
ARE RATED IN AMPS OR
WATTAGE

*Name codes are given in Table 1-5

The cord or wire for the 120-volt condensing section will be #14 SO or #14 SJO. If conduit is being used, #14 TW wire can be used (copper) encased in one-half inch conduit. For a 240-volt condensing section, the cord would be #16 SO or #16 SJO. If conduit and wire is used, #14 TW (copper) and one-half inch conduit should be used. You can not use smaller gauge wire on this application. The smallest gauge wire allowed in general building is #14. If aluminum wiring is used, #12 TW should be used.

Table 1-5. Wire Coding Abbreviations.

NAME	DESCRIPTION	MAX. TEMP.	USAGE
RH	RUBBER COVERED WITH FIBER. HEAT RESISTANT	167 F.	DRY PLACES. NOT USED VERY MUCH. NEED TO SPECIAL ORDER.
RHH	RUBBER COVERED WITH FIBER. HEAT RESISTANT	194 F.	DRY PLACES. NOT USED VERY MUCH. NEED TO SPECIAL ORDER.
RHW	RUBBER COVERED. HEAT AND MOISTURE RESISTANT.	167 F	DRY AND WET PLACES. NOT USED VERY MUCH. HARD TO STRIP AND FISH.
T	THERMOPLASTIC	140 F	DRY PLACES
TW	THERMOPLASTIC, MOISTURE RESISTANT	140 F	DRY AND WET PLACES. MOST COMMON WIRE.
THHŃ	THERMOPLASTIC, HEAT RESISTANT	194 F	DRY PLACES. SKINNY WIRE. USE REWORK.
THW	THERMOPLASTIC, HEAT AND MOISTURE RESISTANT.	167 F	DRY AND WET PLACES
THWN	THERMOPLASTIC, HEAT AND MOISTURE RESISTANT.	167 F	DRY AND WET PLACES. SKINNY WIRE. USED IN REWORKING OLD WORK. MORE CIRCUITS
A	ASBESTOS	329 F	DRY PLACES. STRIP HEATERS. HIGH TEMP.
V	VARNISHED CAMBRIC	185 F	DRY PLACES. MOTORS
AVA	ASBESTOS AND VARNISHED CAMBRIC	230 F	DRY PLACES. STRIP HEATERS. HIGH TEMP.
AVL	ASBESTOS AND VARNISHED CAMBRIC	230 F	DRY PLACES. STRIP HEATERS. HIGH TEMP.
S & SO CORD	RUBBER JACKET WITH PAPER AND JUTE FILLER	SEE LABEL	HEAVY DUTY. DRY AND WET PLACES. MARINE WORK. SMALL COMP. COND. SECTIONS
SJO & SJ CORD	RUBBER JACKET WITH PAPER FILLER	SEE LABEL	LIGHT DUTY . DRY AND WET PLACES. PORTABLE WORK.

ALL GENERAL BUILDING WIRING RATED AT 600 VOLTS
S, SO, SJO, SJ CORD RATED AT 250 or 600 VOLTS. SEE CORD LABEL FOR OTHER INFO.

CHAPTER 2

Test Meters

In meter applications, the same meter movement can be used for the voltmeter, ohmmeter, or ammeter. The difference is, the internal components are arranged and wired in different circuits. The circuit selection switch chooses the mode function of the multimeter and the clamp-on ammeter. Figure 2-1 shows a typical multimeter and clamp-on ammeter.

In Fig. 2-2 the basic circuitry of a multimeter in the ac voltage mode is shown. Resistor A is in series with the meter movement. The very sensitive movement is designed for a maximum load of three (3) volts. The resistor is sized to drop high voltages below the meter limit. The actual voltage read is proportional to the dropping resistor. Most meters have several circuits that can be used at different voltages. They range from zero (0) volts into the thousands of volts.

CHECKING AC VOLTAGE

To check a branch circuit at the entrance panel, remove cover carefully as shown in Fig. 2-3. Be very careful not to cause an arcing by touching the cover against load wires. In Fig. 2-4 you see the procedure. First, remember, electricity travels faster than a blink of an eye. Considering the high cost of multimeters, caution should be used.

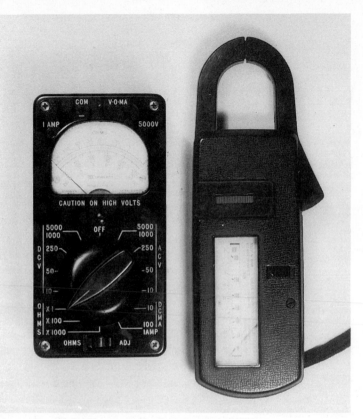

Fig. 2-1. Multimeter and clamp-on ammeter.

■ Always check meter first to see that it is set for ac voltage at its highest scale. A good habit is to place the meter's selection switch to the off position when you finish using it. If it doesn't have an off switch leave it set on the highest ac scale when you put it away.

■ Place meter in parallel with load circuit. Some test leads have alligator clips at their ends; this type requires the load circuit be turned off when hooking up to the circuit. This eliminates the possibility of you receiving an electrical shock. If you have the pin-type leads, the circuit may be left on.

■ Move selector switch to a lower voltage scale until needle deflection is mid-scale on meter face.

NOTE—Many technicians have damaged a meter when they were in a hurry and didn't adhere to this procedure.

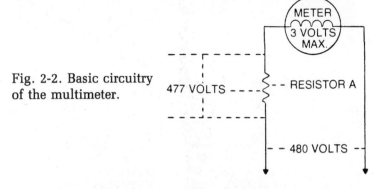

Fig. 2-2. Basic circuitry of the multimeter.

METER
3 VOLTS
MAX.

477 VOLTS - - - - RESISTOR A

- - 480 VOLTS - -

Figure 2-5 shows an ohmmeter with its components. The ohmmeter uses only a small dc voltage through a load and is read when dropped through a resistor. This dropped voltage is equal in ohms resistance dc.

CHECKING RESISTANCE WITH THE OHMMETER

The ohmmeter is usually the second tool used in troubleshooting after determining the correct voltage exists across the load.

All ac or dc loads have resistance. In practical applications, the line voltage to the circuit is turned off. Pull out internal fuses and the ohmmeter should read zero ohms. When testing a coil, or resistor, remove one side of the component electrically from the circuit. Place the ohmmeter across the component and measure the resistance. By disconnecting the component from the circuit, you eliminate the possibility of a voltage from a parallel circuit.

Alternating current coils and motors have less dc resistance than dc coils and motors; the reason for this is that an ac load equals dc ohms resistance plus inductive or capacitive reactance. Reactance is the ability of a load to create its own opposition to current when voltage is applied.

■ A typical multimeter being used as an ohmmeter is shown in Fig. 2-6.
■ IMPORTANT—Make sure there is no line voltage on the load circuit.
■ Place the meter in parallel with the load.
■ Start with the highest resistance range and work down each scale until the needle is close to the center of the meter face, if

Fig. 2-3. Circuit breaker panel with cover removed. Entrance cables are being indicated going into the main circuit breaker.

possible. Check to see if you have continuity to the load ground, case, or shell. When the load you are checking is of low resistance, most of the battery voltage is dropped across the meter drop resistor.

When the load resistance is high, most of the battery voltage is dropped across the load. If you have continuity from the load ground, case, or shell, you have a short circuit.

Fig. 2-4. Meter hookup for testing voltage and amperage.

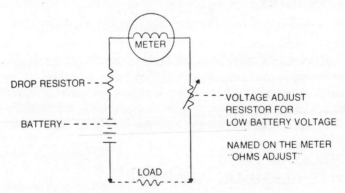

Fig. 2-5. Schematic of typical ohmmeter circuit.

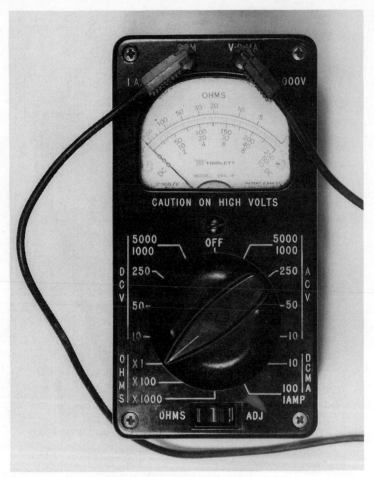

Fig. 2-6. Typical multimeter being used as an ohmmeter.

Remember, there must not be any line voltage connected to the meter. In the ohms mode, the drop resistors' values are too small to conduct line voltage. Some have fuse protection to prevent damage to the meter if you happen to make a mistake by placing line voltage though the meter when in the ohms mode, some do not!

CHECKING CURRENT WITH CLAMP-ON AMMETER

A clamp-on ammeter is used for testing ac only. A single conductor is placed inside the jaw of the meter. With the circuit

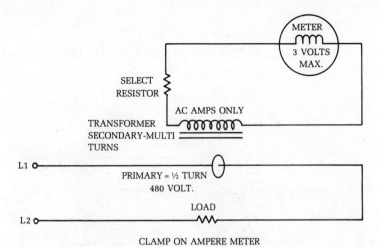

Fig. 2-7. Basic circuitry of a clamp-on ammeter.

loaded, start at the highest amperage scale. Watch the needle deflection and keep lowering the scale until the current reading is near mid-scale. Figure 2-7 shows the basic wiring that accomplishes this operation. Remember the electrical device must be operating and causing a power demand. The power need not be turned off to hook up a clamp-on meter. If bare wires, conductors, or bus bars are being tested, jaws of the meter should be well insulated and extreme caution should be used. In Fig. 2-8 a typical clamp-on ammeter is shown. These meters are important tools in helping a technician to determine if an electrical device is performing according to the manufacturer's specifications. Most manufacturers install a data plate on their equipment that gives all the vital information to the service technician. Usually the first thing listed on the plate is the model and serial number of the device. These become very important at times when a parts replacement might be necessary. Other information given on plate are the voltage and amperage for operating the unit. Abbreviations such as F.L.A. (full load amps) are used on the plate; L.R.A., (locked rotor amps) is a split second current demand to move an armature, or rotor, from inertia into a rotational movement.

CHECKING CURRENT
WITH IN-LINE DC AMMETER

CAUTION—Do not wire an in-line ammeter in parallel with load. The applied line voltage to the low resistance of the meter will cause

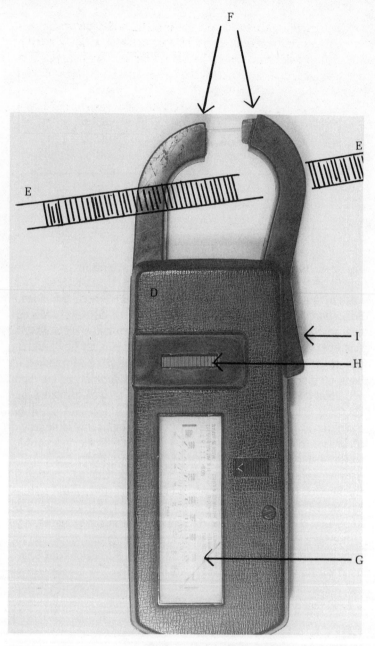

Fig. 2-8. Typical clamp-on ammeter. F—Moveable and fixed jaws. E—Conductor. I—Opens fixed jaw. D—Fixed jaw. G—Meter movement needle. H—Scale selector switch.

very high current flow inside the meter. This will destroy the meter by fire and/or explosion. An ammeter must be wired in **series** with load. The meter should be set at highest scale first, then readjusted. Multimeters such as the one in Fig. 2-1 'A' and 'B' can be used for dc in-line ammeters.

The in-line ammeter has one or more shunt resistors that are rated from nearly zero (0) ohms to five (5) ohms, with high wattage. They are wired in parallel with the meter movement. The meter can be wired in series either before or after the load. The shunt resistor(s) conducts all the current/voltage to the load, except for a small trickle current/voltage that will deflect the ammeter needle. A basic wire diagram explains this in Fig. 2-9. To use an in-line ammeter, the line and load must be de-energized. One line wire must be opened and the meter connected in series with the load at this point.

On dc the meter negative terminal must connect to the most negative side of the circuit, the side with the least potential. On initial start up, use a momentary voltage to observe the needle deflection. If leads are wired backwards, the needle will try to deflect the wrong way, which can result in permanent damage to the meter. If the leads are backwards, reverse the connections to achieve proper meter operation. This only holds true for dc ammeters. The ac ammeters are not sensitive to polarity.

Fig. 2-9. In-line ammeter.

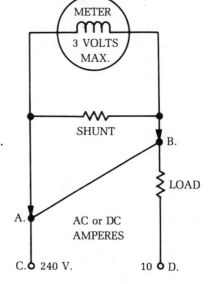

IN LINE AMPERE METER

UNITS AND PREFIXES
FOR ELECTRICAL MEASUREMENT

In the modern system of measurement units, any of the following prefixes can be attached to the basic unit to increase or decrease its size.

Name	Factor	Symbol
milli	0.001	m
kilo	1,000	k
mega	1,000,000	M

For example:

1 milliampere equals 0.001 ampere. In symbol form write 1 mA.

1 kilohm equals 1,000 ohms. In symbol form write 1 kΩ

100 kilowatts equals 100,000 watts. In symbol form write 100 kW.

1 megohm equals 1,000,000 ohms, written symbolically as 1 MΩ

5 megahertz equals 5,000,000 hertz, written symbolically as 5 MHz.

EVALUATION OF ELECTRICAL COMPONENTS

The amount of resistance in dc components can often be calculated by Ohm's Law. Here are some typical components with the resistance you can expect to find in them.

Fuses	zero (0) resistance
Ohmite and wire-wound resistors	Ohm's Law for correct value on ac or dc application. In general, wire-wound resistors should only be used in dc circuits.
Tungsten circuit element	Almost zero (0) ohms when cold. The resistance of tungsten increases with temperature.
Switches	Zero (0) ohms (exception—electronic switches).
Thermistors	zero (0) to two (2) ohms.
Alternating current motors	
Run winding	one (1) ohm
Main winding three (3) phase	one (1) to five (5) ohms

Start windings	five (5) to 40 ohms
Coils on magnetic starters	Readings above 20 ohms
Current coils on relays	less than two (2) ohms
Potential coils on relays	above 15 ohms

Capacitors are used in ac circuits since they do not pass direct current. It is possible to test a capacitor with an ohmmeter using the following procedure.

Set selector switch at highest value. Place probes on capacitor terminals and observe meter needle deflection. The needle will slowly return to infinite resistance. The needle swing and then dropping is due to placing a charge into the capacitor. If there is no deflection of the needle, the capacitor is defective. If the needle deflects but doesn't return to the infinity position, the capacitor is defective. This is covered in more detail in Chapter 4.

Remember when checking the operating current of an electrical device, the amperage should not exceed the limit set by the manufacturer's recommendations found on the data plate.

CHAPTER 3

Compressors

Air conditioning systems that are designed for residential and light commercial use are divided into three sections. Within each section there are many components. The main components are the condensing unit, the evaporator unit, and the thermostat. In some geographic areas, some of the names might be slightly different. For instance, some might call the evaporator unit the air handler. Both are correct.

The condensing units have a specific function, heat rejection. With the rejection of heat many things happen. Figures 3-1 through 3-7 show some condensing units in use today. You can see that they all look different, but they all are designed for the specific task of rejecting heat.

Condensing units are installed wherever the builder thinks it will be cost effective. Because some of you might live in a house, some a condominium, some a duplex, we have written this book for all installation areas. Those of you that can point to your condensing units are lucky. Many owners don't even know where their condensing unit is located. On a service call, you might spend some time trying to locate the customer's unit. It might be under the parking area. Sometimes it is placed on the roof in clusters with other units. This gives you an idea that in certain areas, there might be a little time consumed in finding the condensing unit. When you find it and it is located in a unique place, make a notation on the evaporator unit or entrance panel where the unit is located.

Fig. 3-1. Square condensing unit with vertical discharge.

Condensers can transfer heat using air or water as a transfer medium. Air-cooled condensing units are usually located in the outdoor air to be efficient. Water-cooled units are different. The heat transfer medium can be piped to the unit regardless where it is. A package unit is a unit that has only one section. The condensing unit, evaporator and sometimes the thermostat are located within a single cabinet. I tell you this to save you the embarrassment of looking for an air-cooled condenser on a water-cooled unit.

Before you open the cabinet of the condensing unit, turn off the electrical power supply to it. Always remove the panel slowly for many reasons. You might come face to face with an animal or a

Fig. 3-2. Square condensing unit, ground level on concrete pad.

pressure refrigerant line about to burst. With the panel removed, you can look inside and begin to identify some of the component parts that make the condensing unit operate.

HERMETIC COMPRESSOR

The most expensive part in the condensing unit is the compressor. It is the heart, the pump that circulates the refrigerant as the heart does with the blood. A hermetic compressor is the most common one found in residential and light commercial air conditioning

Fig. 3-3. Typical arrangement of condensing units on ground level in a multi-family structure.

Fig. 3-4. Condensing unit mounted on a platform attached to the structure, where the building code governs flooding areas.

Fig. 3-5. Two condensing units located in a private home on a solid roof area over the screened patio area.

Fig. 3-6. Typical package unit installed on a mobile home. Notice the rainshield that is connected to the home and the unit. This piece of sheet metal protects the ductwork hook-up to unit.

Fig. 3-7. Another typical mobile home installation of a package unit. Notice the window unit that is being used to cool the addition.

and refrigeration systems. In Fig. 3-8 a typical hermetic compressor is illustrated. They are sealed units and cannot be serviced internally in the field. There are re-building shops that have the equipment to cut them open, replace defective parts an weld the shell together again. These compressors are suction cooled. This means that enough cool gas must return to the compressor from the evaporator coil to maintain a desired compressor operating temperature. An electric motor sealed in the shell drives a crankshaft and one or more pistons to operate a reciprocal compressor. In the case of a hermetic rotary type of compressor, vanes similar to those on a water pump impeller or vanes on an oil pump do the pumping instead of the pistons.

Located on the shell are pieces of pipe that have been welded to the body to give access to the high side and low side of the compressor. There might also be small tubes that can be used to charge the system or install pressure operated controls. The larger of the pipes is the low side, or suction side of the system. The smaller pipe is the discharge side, or hot gas line. These short stubs of pipe may be either steel or copper. The refrigeration piping is soldered to them.

Hermetic compressors can be used in small reach in refrigerators and are rated at fractional hp (horsepower). They range from $\frac{1}{16}$ to 60 ton cooling capacity. The larger tonnages are used in

Fig. 3-8. Typical hermetic compressor that is used in most of the residential condensing units.

commercial units. When deciding whether you should have a compressor re-built, first check the price of a new one. Make sure the condition of the rest of the condensing unit warrants an investment as large as a new compressor. If the condenser coil is rotten and the whole cabinet is held together by rust, the customer might want to replace the whole thing. Good customer relations builds confidence in a technician.

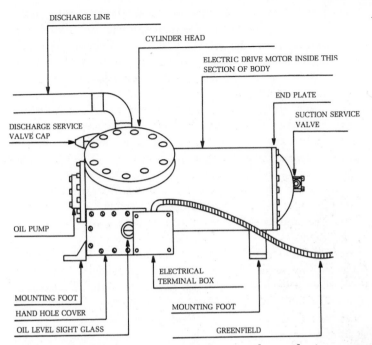

DISCHARGE LINE

CYLINDER HEAD

ELECTRIC DRIVE MOTOR INSIDE THIS
SECTION OF BODY

END PLATE

SUCTION SERVICE
VALVE

DISCHARGE SERVICE
VALVE CAP

OIL PUMP

MOUNTING FOOT

HAND HOLE COVER

OIL LEVEL SIGHT GLASS

ELECTRICAL
TERMINAL BOX

MOUNTING FOOT

GREENFIELD

Fig. 3-9. Semi-hermetic compressor used mostly in commercial equipment.

SEMI-HERMETIC COMPRESSOR

The semi-hermetic compressor is totally different in its construction than the hermetic; this can be seen in Fig. 3-9. The compressor is constructed of a heavy casting. It is bulky and heavy to handle due to the iron content in its body. You will notice that this compressor has nuts and bolts holding it together. This advantage allows the unit to be re-built on a job site. It too has a drive motor (electric) that turns the parts of the reciprocal compressor. Not many of these compressors will be found in residential applications, yet there can be some larger homes or estates that use this kind. Many light commercial applications also use this type of compressor.

Open-Drive Compressor

There are many of these old-timers chugging around the world and still doing the job. Some of you perhaps never saw one of these and never will, but you should know about it in the event you need one for a specific application. These compressors come in a variety of sizes from one (1) hp up. They can be used for air conditioning or refrigeration application. The biggest advantage of this type compressor is the choice of driving power you want. The compressor has a drive shaft protruding from it. The shaft can take either a pulley or a coupler. For instance, I've serviced a unit such as this that was driven by a six-cylinder internal combustion engine fueled by propane. The greatest advantage of this application is the capacity control. With the engine throttle linked to the thermostat, the engine idles when there isn't a load demand. This type compressor can be driven by electric motor, internal combustion engine, or turbine. Another great advantage to this compressor is that it can be installed in an area that doesn't have ample electrical power to drive large compressors.

For all the good, there has to be a little bad. The two biggest drawbacks about the open-drive compressor is its physical size and the critical alignment. As you can see in Fig. 3-10 there has to be

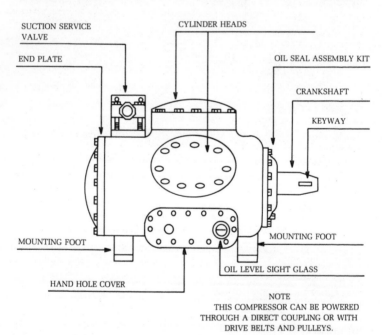

NOTE
THIS COMPRESSOR CAN BE POWERED
THROUGH A DIRECT COUPLING OR WITH
DRIVE BELTS AND PULLEYS.

Fig. 3-10. Open-drive compressor can still be seen in the field today in certain areas.

a close alignment between the compressor and the driving force. If the alignment goes out, beyond specifications, the front shaft oil seal will begin to leak. Both refrigerant and oil will exit here. Alignment with a dial gauge at regular intervals is prudent.

ELECTRICAL CONNECTING TERMINALS

Two types of terminal connections are used with hermetic and semi-hermetic compressors. On the hermetic compressors, pins or terminals are sealed within bakelite or ceramic. The semi-hermetic compressors use terminal boards with threaded bolts being used for the terminals. The board has an 'O' ring on both sides of it. The 'O' ring is designed to compress against the board and prevent the leakage of refrigerant and oil around the terminals.

COMPRESSOR TESTING

With the compressor being the most expensive component in the system, it is wise to be sure it is bad before you condemn it. For this reason, you should learn a systematic method of diagnosing the compressor. You will need a good ohmmeter that can measure from one ohm through 20 ohms. A meter such as this can be purchased in the price range from $20 to $125 depending upon the quality. The homeowner can get by with the less expensive one due to the fact his will not be subjected to the amount of usage the service technician will give his.

This procedure is followed if the compressor doesn't operate when called to do so. Always remember safety comes first. Before opening the condensing unit, turn off the electrical power supply to the unit. With the service panel removed, look with your eyes before you touch anything. There should be some type of flash cover enclosing the terminals of the compressor. This is a basic safety device to protect the service technician from electrical shock when the unit is operating, and it protects the technician from pressure-driven oil if one of the terminals should fail and blow out of its mount. Don't assume that oil in a unit is clean. In most units the refrigeration oil is clean; however, in some instances, the oil has become contaminated. The formation of sulfurous acid might have occurred inside the inoperable unit. This acid can be dangerous to your skin and eyes. For this reason, don't assume anything; be sure and careful. With the cover removed, you will notice that the terminals are arranged in the approximate order as shown in Fig. 3-11. The

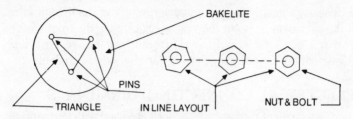

Fig. 3-11. Typical compressor terminal connections arrangement.

following is a list of electrical failures that you will test for.

a—Grounded compressor (short circuit). This condition takes place when the insulation of the drive motor windings leak the electricity to the steel compressor body. Blown fuses result.

b—Open-winding. A condition that occurs when the conductor of one of the motor windings parts.

c—Locked rotor. This condition happens when either the crankshaft bearings seize due to lack of lubrication, or a compression component breaks within the compressor shell jamming the crankshaft. In the case of a single-phase unit, the same locked-rotor condition will be witnessed if the system has a defective starting component.

The multimeter enables us to diagnose internal electrical problems of the compressor. "Ringing out a compressor" means taking continuity tests. Make sure the power is off. If necessary, turn your meter to ac volts and check it out. Sometimes a disconnect leaves a blade engaged that has broken loose from the main control bar. After you check for voltage, mark the wires that are connected to the terminals so they may be returned to the same position when re-assembled. There are many ways to mark them: different color tape, black bands of electrician's tape, notch with knife blade. Whatever works for you is suitable. The wires must be removed from the compressor to prevent voltage from other circuits. For instance, a 240-volt compressor might have a 120-volt condensing fan motor. It is possible that you might read the neutral leg of the 120-volt circuit as a grounded compressor.

Zero your meter. Make sure that it rests at infinity. This is done with the little screw at the base of the indicator needle. Then touch the two probes together and turn selector switch to X1000 scale. The needle should deflect to zero ohms. If not, make adjustments with the little knob to set your meter to zero.

Testing for Grounded Compressor

Scrape the copper suction line on the compressor so it shines. This can be done with a knife blade or a piece of sand cloth. Place one probe on the clean copper surface, and the other one on a compressor terminal. Check each one by moving the probe to each terminal. If there is no deflection on the meter, the compressor is not grounded. If you have a reading, the compressor is bad and you need not check any further. The three terminals are electrically connected internally.

Testing for Open Windings

Place a test probe on terminal one. With the other probe touch terminal two. The meter needle should deflect. This shows there is a circuit. Repeat the procedure until the circuitry between the terminals is confirmed. There should be a reading across each pair of terminals.

The windings in a three-phase compressor are different than a single-phase compressor. In the three-phase compressor, you should read the same amount of resistance through the three windings. This is not true in the single-phase compressors. The reason is that the start-winding has more wiring turns in order to develop more torque on start up. The run-winding has a heavier gauge wire with fewer turns, thus the resistance readings will be different between the three windings. Knowing this, it is possible to identify the windings in a single-phase compressor.

RUN—The lowest reading of resistance—(about one ohm)
START—The middle reading of resistance—(about five to 22 ohms)
COMMON—The maximum resistance reading—(total of all windings)

Rotation

The single-phase hermetic compressor has a fixed rotation of its electric motor. In three-phase applications rotation is very important. The compressor might have a directional oil pump that will not pump efficiently if it turns in the wrong direction. Rotation on three-phase motors may be reversed by reversing any two motor leads.

This can be done either at the motor starter or at the motor itself. It is easier to do at the starter or disconnect most of the time. Be careful not to cause trouble in another circuit or cause a cross-phasing. This occurs when you touch two phases together without

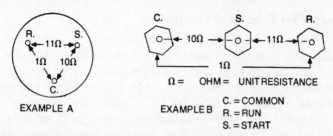

Ω = OHM = UNIT RESISTANCE

C. = COMMON
R. = RUN
S. = START

EXAMPLE A   EXAMPLE B

Fig. 3-12. Example A shows approximate resistance readings when a compressor is being rung out. Example B shows the same readings with a different terminal arrangement.

a load circuit in between. Figure 3-12 shows a compressor being rung out, note the readings at the terminals.

Locked Rotor Test

The compressor hums but will not start. The overload relay usually opens the common winding either internally or externally, this allows the windings to cool and not get hot enough to melt. This is very apparent when you place your hand on the compressor shell. It is very hot to the touch, and it would be difficult to keep your hand on the compressor.

■ Turn off the power to the compressor.
■ Remove all extra machine wiring attached to the compressor motor terminals.
■ Ring out the compressor and label the common, start, and run pins. (single phase)
■ Secure line one to the run pin and line two to the common pin.
■ Place insulated jumper wire from run to start. (see Fig. 3-13)
■ Turn on the power.
■ If the locked rotor breaks lose, the compressor will start and come up to speed in less than five seconds.
■ After compressor reaches full speed, remove jumper wire with compressor operating, leaving line one on the run pin and line two on the common pin.

To start a hermetic compressor motor which has a start capacitor, repeat the first four steps above. Step five is to place jumper wires between the run and start capacitors and from the start terminal to the start capacitor. It doesn't make a difference on the hook-up of the jumper wires to the start capacitor terminals. See

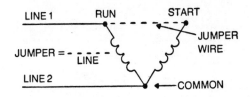

COMPRESSOR
INDUCTION START; INDUCTION RUN

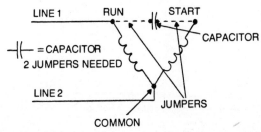

Fig. 3-13. Capacitor start hermetic compressor motor diagram.

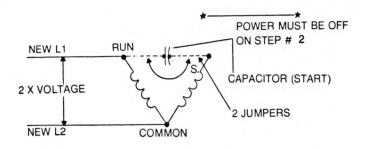

Fig. 3-14. Repeat steps six and eight for induction-start, induction-run motor. In step eight remove jumper wires and start capacitor with the compressor operating. Line one to run and line two to common remain (L1 and L2).

Tables 3-1 through 3-3 show operating current information. With your ammeter you can determine if a specific horsepower motor is operating properly. The tables are also helpful in sizing overloads and heaters for motor starters, especially when the data plate on the machine is missing or not legible. Never exceed the rated amperage of a motor. If you do, the motor will have a short life due

Table 3-1. Operating Currents
for Three-Phase AC Motors at Full Load.

HP	240 V	480 V	2300V
AC-MOTORS FULL LOAD.			
½	1.91	.96	
¾	2.68	1.34	
1	3.45	1.73	
1½	4.98	2.49	
2	6.51	3.26	
3	9.19	4.6	
5	14.55	7.28	
7½	21.05	10.53	
10	26.8	13.4	
15	40.2	20.1	
20	51.68	25.84	
25	65.08	32.54	
30	76.56	38.28	
40	99.53	49.77	
50'✱	124.41	62.21	
75	183.74	91.87	20
100	237.34	118.67	26
60	147.38	73.69	16

Table 3-2. Operating Currents
for Single-Phase AC Motors at Full Load.

HP	120	240
1/6	4.22	2.11
1/4	5.56	2.78
1/3	6.9	3.45
1/2	9.39	4.69
3/4	13.22	6.61
1	15.33	7.66
1 1/2	19.16	9.58
2	23	11.5
3	32.57	16.29
5	53.65	26.83
7 1/2	76.64	38.32
10	95.81	47.9

to overheating the windings. Always remember that these ratings are given to be the maximum, when the unit has the maximum load on it, whether the motor drives a fan or a pump. All amperage ratings are given for the maximum. For instance, you are topping the charge of a reach-in freezer, or a walk-in freezer that is operating at the

Table 3-3. Locked Rotor Amperages.

HP	SINGLE PHASE		THREE PHASE	
	120 V	240 V	240 V	480 V
½	56.33	28.17	11.49	5.75
¾	79.32	39.66	16.08	8.04
1	91.97	45.98	20.1	10.05
1½	114.96	57 48	28.71	14.36
2	137.95	68.98	37.32	18.66
3	195.43	97.91	51.68	25.84
5	421.89	160.94	86.13	43.07
7½	459.84	229.92	126.32	63.16
10	574.8	287.4	155.08	77.54
15			229.68	114.84
20			298.58	149.29
25			367.49	183.75
30			447.88	223.94
40			597.17	298.59
50			717.75	358.88
60			861.3	430.65
75			1062.27	531.14
100			1412.53	706.27
125			1780	890

time of charging at five degrees below zero (−5 degrees F.), if you bring the compressor up to maximum amperage at this time, the unit will draw excessive amperage when terminating its defrost cycle and entering the freeze cycle. In fact, the compressor might trip its thermo-overload at that time.

Hot-Shotting the Compressor

This is another method that can be used to try to break loose a locked rotor in a single-phase compressor.

■ Make sure the start and run capacitors have high enough ac voltage rating for the new applied voltage.

■ Remove wiring from the compressor motor terminals. Double the line voltage hook-up as in Fig. 3-14.

■ Make sure the power is off while you are doing the second step.

■ If the compressor is 120 volts ac, make the line one to number two 240 volts ac. If the compressor is rated at 240 volts ac, single-phase, make line one to number two 480 volts ac, single-phase.

■ Attach a jumper from run to capacitor and a second jumper from start to capacitor.

■ Take the jumper wire off the start terminal of the compressor.

■ Turn on the higher voltage.

■ Take the jumper wire and tap it about four times (one second each, to the start terminal). Do not touch the live voltage. Be careful and hold the insulation of the jumper wire.

■ Turn off the power, then repeat the above procedure in five minutes.

Reverse Rotation Method

■ Turn off power and remove the wiring from the hermetic compressor motor terminals.

■ Study Fig. 3-15.

■ Wire line one to common and line two to start. Use rated compressor line voltage.

■ Attach the jumper wire from run to capacitor and second jumper wire from start to capacitor.

■ Disconnect jumper from motor run terminal.

■ Turn on the line voltage and hold jumper wire by the insulation; then hold jumper to run for four seconds, four times at four-second intervals.

■ If motor does not reverse, repeat step six using 240 volts ac instead of 120 volts ac motor rated voltage. Use 480 volts ac instead of 240 volts ac for a motor rated at the higher voltage (single-phase).

Check the capacitor for a higher rated voltage. When you have completed steps one through seven using double line voltage, you

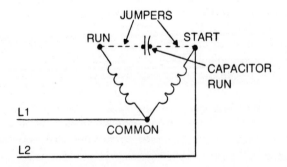

Fig. 3-15. Reverse rotation method to break a locked rotor.

have hot-shotted the compressor in reverse. Do not hot-shot in reverse until you have done steps one through seven with rated line voltage for the compressor. You might be able to break the locked rotor with normal line voltage. There is no need to strain the compressor motor unless absolutely necessary. If these steps don't free the locked rotor, there is nothing else you can do in the field.

That covers the important electrical problems that you will find in the field with a compressor. The other malfunction of the compressor is mechanical failure.

REFRIGERATION VALVES
(COMPRESSOR REED VALVES)

When the compressor is in a refrigeration system and in operation, if the suction-side and high-side pressures are almost the same, the valves might be bad. An ammeter will tell you if the electric drive motor is doing the work for which it is rated. Without proper compression in the compressor, the amperage reading will be very low. The problem could be a piston not pumping. When a compressor is operating properly, it should draw an amperage close to its rated amount on the data plate. The suction line should be cool to the touch and sweating. This cool gas is needed to help cool the compressor windings. The small liquid line should be warm to the touch . . . not hot. If the liquid line is very hot when in the cooling mode, there is a problem with the unit. Remember, refrigeration equipment and air conditioning equipment have design temperatures and conditions. In different areas of the world there will be different designs in the equipment. In a low humidity area, the suction line might not sweat. If a unit is designed to attain 74 degrees F. conditioned space with an ambient of 95 degrees F., and it is being checked on a day when the ambient is 80 degrees F., the equipment will accomplish the 74 degree F. without any problem. Even if there was an inefficient compressor with a possible bad valve, it wouldn't show that easily until the unit was being operated under its design conditions. Conditions such as the evaporator coil being clean or the condenser coil blocked with grass cuttings will affect the operation of the entire system. Another thing that I learned a long time ago, know what the engineer wanted the equipment to do. If you don't know what it is supposed to do, how can you repair it? As a service technician, you will see many applications of the refrigeration theory, from food processing to industrial manufacturing. That is why it is important that you know what the unit is supposed to do before trying to make a repair.

CHAPTER 4

Capacitors and Fuses

Capacitors are used on equipment with single-phase electric motors powered by alternating current. The five major types of electric motors are:

Shaded-pole motors. Small induction motors use shaded poles for the purpose of starting. They have very low starting torque and are used to drive a fan directly. These motors are available in fractional horsepower sizes.

Split-phase motors. These are induction motors with separate windings for starting. The start windings are removed from the electrical circuit with the use of a centrifugal switch when the full running speed is approached. The split-phase motor still doesn't have enough starting torque to be used with heavy loads. They are in the fractional horsepower range and may be used to power direct-drive fans or small belted fans.

Capacitor-start motors. These induction motors have a separate start winding with a capacitor (electrical condenser) connected in the start winding for added torque. When the motor approaches full running speed, the start winding is disconnected and the motor operates as a straight induction motor. These motors have a heavy current demand upon starting. They too are sized mostly in fractional horsepowers.

Capacitor-start and run motors. This motor is similar to the capacitor-start motor except that the capacitor and starting windings are designed to remain in the circuit thus eliminating the switch to disconnect the starting windings.

Repulsion induction motors. They were the work horses of the single-phase alternating current line. These motors are high torque and are used to drive larger equipment.

TYPES OF MOTOR CAPACITORS

Start Capacitors

These capacitors are usually cylindrical in shape and have a microfarad rating from 90 to 400. There are variations of this range from time to time. The plastic outer case encloses alternate layers of paper and a dry electrolyte. These capacitors are designed to stay in the circuit momentarily and are removed electrically by means of a transfer switch. This is accomplished with the use of a relay or a centrifugal switch. The start capacitor should remain in the circuit only long enough to get the motor up to two-thirds to three-quarters of its rated rpm. If it is not removed from the electrical circuit at that time, the capacitor would rupture, burst, or explode due to overheating.

Run Capacitors

The run capacitor is a constant-duty capacitor. This means that it is designed to stay in the circuit electrically while the motor is in operation. The run capacitor is the direct opposite of the start capacitor. The run capacitor has high impedance and low capacitance, for this reason they are usually rated between five and 40 microfarads. Their construction enables them to dissipate heat. With the oil-filled metal body, this capacitor can stay in the circuit without the danger of bursting. There are times when the capacitor will burst, and it will spew oil all over its general location. Figure 4-1 shows a typical run capacitor.

When selecting the voltage ratings of motor capacitors, choose ones with a working voltage at least equal to the operating voltage of the unit or more. For instance, when replacing a run capacitor on a unit rated to operate at 240 volts ac, capacitors with voltage ratings of 370 or 440 volts ac can be used. In fact, it is sometimes difficult getting the exact voltage rating on a capacitor. Don't replace a 220-volt ac capacitor with one rated for 110 volts ac. Remember a higher voltage rating is alright; using an underrated capacitor may be dangerous. In Fig. 4-2 three types of capacitors are shown. It also shows the multimeter hook-up for the test below.

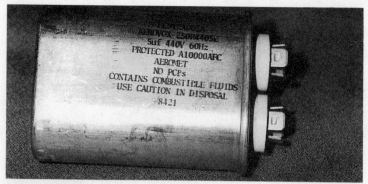

Fig. 4-1. Typical run capacitor.

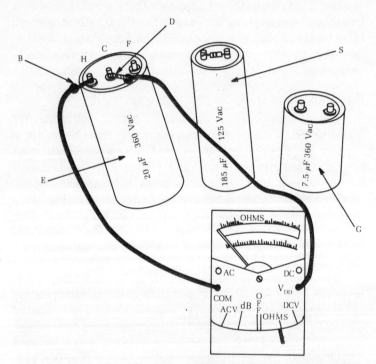

Fig. 4-2. Run capacitor E is a dual capacitor. Terminal C is the common and H is hooked to the compressor, while F is wired to the fan motor. This capacitor is very common in window units. Capacitor S is a dry electrolyte start capacitor. The small capacitor G is a fan capacitor.

CHECKING A CAPACITOR

■ When checking capacitors, the first step is to set your multimeter selection switch to the highest resistance scale (ohms).

■ Check your meter to zero and infinity.

■ Always discharge the capacitor that is to be checked before handling. If you don't and the capacitor has a charge in it, you will get quite a shock. The recommended way of discharging a capacitor is to place a 15,000 ohm, two-watt resistor across the capacitor terminals. Some people use a small piece of insulated wire, other just place the blade of a screwdriver across the terminals. The resistor is the best method to use, especially with those capacitors that have an internal fuse. Sometimes shorting with a screwdriver can cause this fuse to fail rendering the capacitor useless. With the capacitor discharged, you must remove one side of the shunt resistor from one of the terminals.

■ Place one probe of your multimeter on one terminal. Now place the other probe on the opposite terminal. The meter needle should deflect fully, then in a split second start returning to infinity. This is caused by the small dc battery in the meter placing a charge in the capacitor. If you want to repeat the test or if it didn't work, reverse the probe positions which will reverse the polarity. If the needle reacts in the described way, the capacitor is serviceable. If not, the capacitor will need to be replaced. If there is a slight deflection of the needle and it doesn't return to the infinity position, the capacitor is shorted and should be replaced. With metal case capacitors, check continuity from the terminals of the capacitor to the case. There should be absolutely no reading. If there is a slight deflection, replace the capacitor.

Run capacitors usually have an index mark on their tops, something that designates where the line wire should be placed to charge the capacitor. It is sometimes a red dot, an index bump, or a plus sign. It is true that the capacitor will function whether the line is indexed or not, but for safety reasons the line wire that supplies the voltage to the capacitor should be indexed. The reason is the way the capacitor is constructed. The terminal indexed is electrically attached in the center of the metal casing. The other terminal is wired very close to the inside of the metal shell. If for any reason the wire should come loose inside the capacitor, it could blow a hole in the case spewing hot oil. It's simple enough to wire the line to the indexed terminal. Figure 4-3 shows capacitor mounting in a General Electric condensing unit.

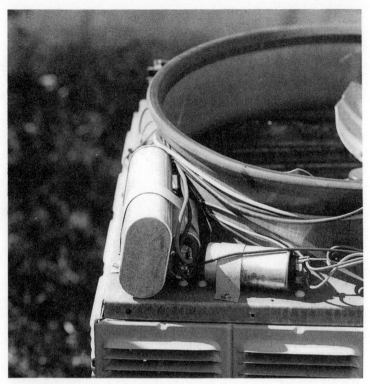

Fig. 4-3. Actual mounting arrangement of capacitors in a General Electric condensing unit.

HOLDING-A-CHARGE TEST

When checking resistance in the capacitor, you already know that you've placed a charge in the capacitor from the meter battery. To check if the capacitor retains this charge, turn the selector switch of your multimeter to the dc voltage scale for 10 volts. Touch the probe to the terminals of the capacitor; the needle should deflect to about the 10 volt mark. If the needle deflects towards infinity, reverse the probes to correct the polarity.

Those of you that have been in the field a while are very familiar with what happens to the identification markings of the capacitors once they have shed their packaging and begin sliding around the shelves in the back of a service truck. After a short time, they become unreadable. Knowing the frustration of this experience, you will all take extreme caution in keeping your capacitors pampered. On the other hand you might know all the numbers on your capacitors, but the piece of equipment doesn't have any identification of the capacitors that need replacement. For this reason, two tables have been placed in this book showing capacitor ratings for fractional horsepower compressors (see Tables 4-1 and 4-2). If you have a compressor that won't start but has indications of locked rotor, you might want to check the capacitors as described in this chapter.

Due to the fact that a start capacitor is only in the circuit electrically for just an instant of time, it could be replaced with a capacitor of a slight difference in microfarad rating. The run capacitor is more critical, and the same size should be used for replacement. There should not be a hum in the motor or the compressor when operating. The run current should also be watched for excessive amount over factory specifications. If the compressor starts alright and then develops a hum, it is possible that the run capacitor has a too high a value. Lower it and check it again.

Three phase (polyphase) motors do not start components, they are only used on single-phase equipment. Many units are using a P.S.C. (permanent split capacitor) compressor, this means there is only one capacitor used in the unit. The disadvantage is that the compressor will not start under a load. The single capacitor requires that the unit shut down for several minutes before trying to restart.

Table 4-1. Run Capacitors
Approximate Sizing for Horsepower of Compressor.

mF stands for microfarads

COMPRESSOR or motor	⅛ HORSEPOWER	CAPACITOR WILL BE NEAR 4 to 5mF
COMPRESSOR or motor	⅙ HORSEPOWER	CAPACITOR WILL BE NEAR 4 to 5mF
COMPRESSOR or motor	½ HORSEPOWER	CAPACITOR WILL BE NEAR 10
COMPRESSOR or motor	½ or 2 HORSEPOWER	CAPACITOR WILL BE NEAR 10 to 15mF
COMPRESSOR or motor	3 HORSEPOWER	CAPACITOR WILL BE NEAR 2 capacitors at 10mF. Total 20mF
COMPRESSOR or more	5 HORSEPOWER or more	CAPACITOR WILL BE NEAR 2 capacitors not to exceed 40 microfarads.

Table 4-2. Start Capacitors
Approximate Sizing for Horsepower of Compressor.

mF STANDS FOR MICROFARADS

COMPRESSOR ⅛ HORSEPOWER	CAPACITOR WILL BE NEAR 95 TO 200μF
COMPRESSOR ⅙ HORSEPOWER	CAPACITOR WILL BE NEAR 95 TO 200μF
COMPRESSOR ¼ HORSEPOWER	CAPACITOR WILL BE NEAR 200 TO 300μF
COMPRESSOR ⅓ HORSEPOWER	CAPACITOR WILL BE NEAR 250 TO 350μF
COMPRESSOR ½ HORSEPOWER	CAPACITOR WILL BE NEAR 300 TO 400μF
COMPRESSOR ¾ HORSEPOWER	CAPACITOR WILL BE NEAR 300 TO 400μF

If it short cycles, it kicks out the circuit breaker, or the overload might open, or even blow a fuse. In certain regions, power supply demand is so high that the power companies are not supplying full power ratings. It is this reason that a 10% electrical design fluctuation is built into the equipment. When power supply borders on this 10%, problems can be witnessed with hard starting. Hard start kits have been installed in many areas. The kit consists of a start capacitor and relay, and at times a run capacitor. In extreme cases of short cycling, a time delay is used to keep the unit off until the pressure in the system equalizes.

FUSES

Too many people think fuses are all alike. In fact some will even say, "a fuse is a fuse," but that is not true. There are many types of fuses, and each of them is designed for a specific purpose. The most important reason for using the proper fuse is to have a device that will melt and open a circuit before sufficient heat is generated that could melt the conductors and perhaps ignite the entire structure. For this reason, three factors are very important at this time. Wire size, load amperage, and fuse size. They are closely related. The fuse sizing must be below wire ampacity. For the protection of the appliance, the fuse should be rated close to the f.l.a. of the appliance. In Fig. 4-4 some of the different types of fuses are shown. The job of the fuse is to open the circuit when a load exceeds the rating of the fuse.

The glass screw type fuse is still in service in many regions. The biggest problem with this type of fuse was that the base of it was the same for all the different ampacities. If a homeowner plugged an appliance into a wall receptacle, and it caused a circuit overload, the fuse would open. If the fuse that blew (opened) was rated for 15 amps, and all the owner had was a 20 or 30 amp replacement, he would use it. I'm sure this practice caused many structural fires or wiring damage.

The fuse-stat type was then introduced. Basically the difference between it and its predecessor was the threading on the screw-in base. Each fuse-stat had its own color that indicated its ampacity. The safety feature was that each ampacity fuse had a different threading on its base. This meant, if a 15 amp (blue) fuse was to be changed, nothing but a 15-amp fuse could be screwed into the socket (fuse holder). There are many areas still using this type of fusing.

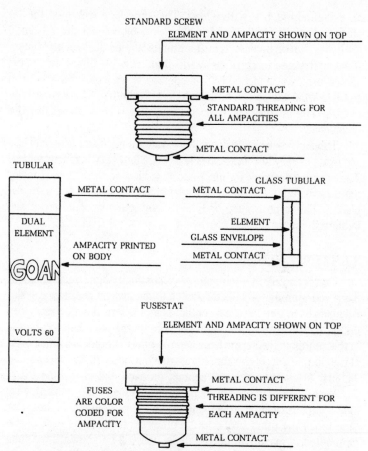

STANDARD SCREW

ELEMENT AND AMPACITY SHOWN ON TOP

METAL CONTACT

STANDARD THREADING FOR ALL AMPACITIES

METAL CONTACT

TUBULAR

METAL CONTACT

GLASS TUBULAR

METAL CONTACT

DUAL ELEMENT

ELEMENT

GLASS ENVELOPE

AMPACITY PRINTED ON BODY

METAL CONTACT

GOAM

VOLTS 60

FUSESTAT

ELEMENT AND AMPACITY SHOWN ON TOP

METAL CONTACT

FUSES ARE COLOR CODED FOR AMPACITY

THREADING IS DIFFERENT FOR EACH AMPACITY

METAL CONTACT

Fig. 4-4. Different types of fuses used in air conditioning field.

Newer homes use circuit breakers. This type of protection was widely accepted due to the fact there was no longer a need to find an open hardware store late at night when a fuse would fail. With the circuit breaker, the homeowner waits until it cools, turns it to the off position and then to the on position, and he has service again. If the breaker trips again, he knows there is a problem on the circuit that will need correction. Some residential-type circuit breakers can be reset in a single motion.

In both fuses and circuit breakers there are two types of elements placed inside of them. One is called a quick-blow element and the other a slow-blow element. Remember this difference, for it might save you a lot of headaches someday. The quick-blow fuse is designed for the lighting circuits or any circuit that doesn't carry an extremely high resistive load. This type fuse will open the instant the ampacity is exceeded. The slow-blow fuses are designed for heavy starting loads. It is sometimes labeled dual-element fuse. It is constructed to carry an overload for a couple of seconds. This allows it to carry a locked rotor amperage when a motor starts. You can see what havoc can be initiated, with the use of a wrong type of fuse. The mistake can be made by both the homeowner and the technician.

Tubular fuses are designed for a wide range of ampacities. They are made in very low ampacities and range into very high ampacities. They too are designed for quick and slow-blow service. Some tubular fuses have removable ends that enable the element or link to be replaced. This type of fuse body is usually found in commercial and industrial applications.

TESTING FUSES

To test a fuse, the ohmmeter or continuity light must be used. Turn your ohmmeter selector switch to the highest resistance scale. All fuses have two external contact points where the electricity can enter, then exit the fuse. With the screw type fuse, one contact is in the center of the screw base and the other contact is a small metal tab on the side of the threading where it butts to the receptacle. Tubular fuses have a metal contact on each end of the fuse.

With the ohmmeter on highest resistance scale, or continuity light switch turned on, touch one probe to one contact, at the same time touch the other probe to the other contact. In the case of the ohmmeter, a maximum needle deflection will be shown if the fuse is serviceable. The continuity light will illuminate if the fuse is serviceable. If nothing happens the fuse is open and should be replaced. If there is a slight deflection of the needle, the fuse should be replaced.

CHAPTER 5

Relays and Contactors

In Chapter 4 different types of motors were discussed. The centrifugal switch was discussed to remove starting components from the circuit when the motor reaches a specific speed. This method cannot be used in a hermetic compressor. Specially designed relays were designed to remove the starting components electrically from the circuit when the compressor reaches almost full speed.

CURRENT RELAY

All relays, with the exception of electronic ones, have an armature of iron core with a conductor wrapped around it with a specific amount of turns. When electricity flows through the wire, a magnetic field is developed in the iron core. This is the way a switching action is accomplished. The current relay is designed to activate upon high current demand. The switch is in the open position when no current is applied, this is n.o. (normally open). When a high current is placed through the windings, the switch immediately closes allowing the starting components to be placed into the circuit with the starting windings. As the speed of the compressor drive motor increases, the current drops. This allows a small spring to apply pressure on the moveable contacts and open the switch. The one disadvantage of this type of relay is the current surge across the contacts causes excessive arcing that deteriorates the contacts. These relays are rated by current values which differ with the

Fig. 5-1. Typical relay with terminal arrangement. This unit has dpdt (double-pole, double-throw switching action).

horsepower of the compressor to be controlled. Figure 5-1 shows the typical terminal layout. The internal parts of this relay are shown in Fig. 5-2.

POTENTIAL RELAY

This relay is similar in appearance and construction to the current relay. The main difference is that the contacts are normally closed thus eliminating much of the arcing that is exhibited by the current relay. This is accomplished by using a coil-wound armature that is magnetized when high voltage is applied. The coil is series-wired to the common winding of the compressor motor. As the compressor starts, supply voltage drops. This condition permits the contacts to remain closed leaving the start capacitor electrically connected in the circuit. When the compressor reaches about 60% of its full speed the voltage also rises. This higher voltage causes the armature to

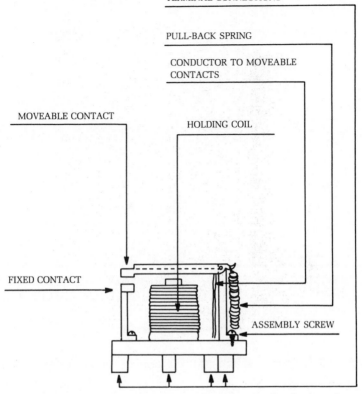

TERMINAL CONNECTIONS

PULL-BACK SPRING

CONDUCTOR TO MOVEABLE
CONTACTS

MOVEABLE CONTACT

HOLDING COIL

FIXED CONTACT

ASSEMBLY SCREW

Fig. 5-2. Simple workings of a single-pole, single-throw relay.

open the contacts that remove the starting capacitor electrically from the circuit. These two relays look alike at first glance. When replacement is needed, examine them very carefully to determine which type relay is needed. Figure 5-3 shows how the potential relay is wired into a circuit.

HOT-WIRE RELAY

This relay works on an entirely different principle. Note Fig. 5-4 and you will immediately see the difference. The hot wire relay has bimetal contacts. Located close to the bimetal is a chrome nickel high resistance wire. Without an applied voltage, the contacts remain in a closed position. When voltage is applied, it flows through the chrome nickel wire causing it to heat. It is designed to each maximum heat when the compressor reaches full operating speed, at which

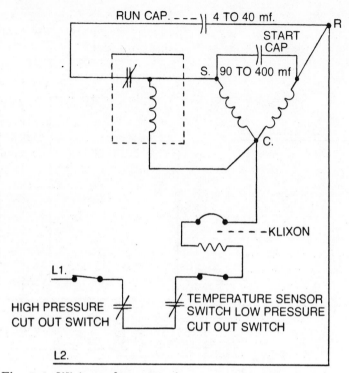

Fig. 5-3. Wiring schematic of a potential relay in a circuit.

time the first set of bimetal contacts open. When these contacts open, the starting capacitor is removed electrically from the circuit. If the compressor is not operating properly and causing an over-heating condition, a second set of bimetal contacts opens and causes the compressor to stop. With the compressor shut down it has time to

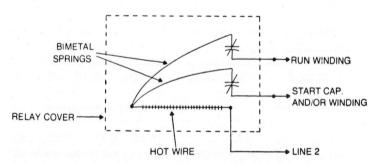

Fig. 5-4. Wiring schematic of a hot-wire relay.

60

cool. This condition also allows the bimetal contacts to cool and reset in approximately four minutes. This type of starting relay also gives the additional protection of the thermo-overload.

BIMETAL CONTACTS

Some homeowners might be scratching their heads at this point as to the meaning of bimetal. Let me clear up the mystery of the meaning of this word. Different types of metals expand (stretch) and and contract (shrink) with heat and cooling at different rates. For example, at the same temperature copper wire and aluminum wire expand at different rates. If the two wires were made into flat pieces of metal and bonded together, you would have bimetal strip, made of two dissimilar pieces of metal. With one side expanding at a faster rate than the other side, the strip will bend. This is how the switch action is accomplished with a bimetal switch. This is exactly how a klixon motor overload operates. The factor that determines when the switch actuates is the heater. The resistance factor in the material used for the heater determines the amount of heat generated. The material will allow electricity to flow until the current causes a heating effect exceeding the designed amount. This action is similar to what happens to a fuse when its rating is exceeded.

RELAYS AND CONTACTORS

The relays discussed in this chapter were designed for the specific job of starting a compressor motor. They provide an automatic way to place a start capacitor into a circuit electrically and remove it the same way. There are many different types of relays, but they are all switches that are activated by an electro-magnetic force, or by heat. Most relays are designed to carry very light current loads. They average from a few amps up to about 15 amp loads. Many times in a circuit, a relay actuates a device that is capable of conducting a larger ampacity load. The holding coils of the relays can be wired to conduct 24, 120 or 240 volts. When using a relay, be sure you know the operating voltage in the circuit where it is to be installed.

Contactors

Contactors are used as electrical switches that can conduct high ampacity across their contacts. The larger the load becomes, the larger the physical size of the contactor becomes. The contactor can

Fig. 5-5. A common relay.

also be wired with a holding coil that will operate with 24, 115, 230 and sometimes higher voltages. In Fig. 5-5 a relay is shown; compare with the contactor shown in Fig. 5-6 and you will definitely be able to see the difference in their construction and size.

Fig. 5-6. Contactor.

Contactors and relays are electrically rated the same way. First is the ampacity rating, which is the amount of amperage each set of contacts can conduct safely without causing damage to themselves. Some have silver-plated contacts that begin to melt when load is exceeded. For this reason stay below the rated amperage for the load it is to carry. Second, is the voltage rating for the coil; this is the amount of voltage that has to be supplied to the armature in order to set up a magnetic field. Third, the number of circuits that can be wired through the contacts. In the case of a contactor, you might have a single-pole, double-pole, three-pole, or a four-pole switch. In a relay the same rule applies. This is how an abbreviation might be shown on a relay, d.p.s.t.n.o. (double-pole, single-throw, normally-open). The position of the contacts are given in the de-energized position of the relay. It is either n.o. (normally open) or n.c. (normally closed). The number of poles on the relay will depend upon how many functions you want the relay to perform.

In most residential systems you will only be dealing with line voltage which can be 120 volts and/or 240 volts with a control voltage of 24 volts. In commercial units voltage can go higher in both the line and control circuits.

CHAPTER 6

Transformers

In Chapter 1, transformers used by the power companies were discussed. Transformers are used either to reduce or to raise voltages to become more workable. There are many types and sizes of transformers, but they all basically do the same thing. In this chapter only those transformers used in residential and light commercial units will be covered.

The transformer shown in Fig. 6-1 is the most common transformer in use today; it is found in residential and light commercial units. Within the housing of the transformer used in residential units, there are two windings. The first is called the primary, and the second is called the secondary. The line voltage, being the primary, enters the primary winding side of the transformer. The step-down voltage, usually being 24 volts ac, is taken from the secondary winding terminals. Some of the newer residential air conditioning transformers are capable of using either 120, 208, or 240 volts ac and still have a secondary voltage of 24 volts ac. This is accomplished with having the primary winding tapped to the outside of the transformer at different lengths of the winding. When installing a transformer in a unit with a transformer with multiple primary voltages, use your voltmeter to measure the actual input voltage. Many units might have a data plate stating the unit operating voltage, this might not be the actual voltage supplied by the power company lines. Remember, the primary must have the rated amount of voltage to produce the 24 volts on the secondary. If there is a

Fig. 6-1. Typical step-down transformer used in air conditioning.

low voltage problem on the primary side of the transformer relays and contactors might not operate properly causing many problems in the control circuit.

WATTAGE RATING

What is VA? It is the abbreviation for volt-amps. This is another way to say watts. The average household light bulb is rated in watts. Appliances are also rated in watts. Transformers are rated in watts. When asking for a transformer from a parts store, they will ask for the wattage. Why? First remember how to find watts.

$$V \times A = W$$

The computation is also done for the secondary side of the transformer.

$$24 \times 2.08 = 49.92 \text{ watts}$$

This would be close enough to consider 50 watts. This would be a 50 watt transformer. The more common is the 40 watts transformer.

$$24 \times 1.66 = 39.84 \text{ watts}$$

Why do you have to know this? An electro-magnetic coil requires energy to operate. The amount of electrical power needed to magnetize the iron core is calculated in VA. The electro-magnetic coil is also called the holding coil. In each unit there are a certain number of holding coils in the control circuit. The rating of these holding coils in relays and contactors might be expressed in amperes or VA. If given in amperes, you will have to figure the entire load on the transformer. For example, if there are three holding coils, and each of them draws 0.3 amps, the total draw, if the three were being energized at the same time, would be 0.9 amps. You can see that a 40-watt transformer will supply 1.66 amps without a problem. Many manufacturers are now rating their holding coils in VA. This means you just have to add the total VA consumption of all the holding coils that will be energized at the same time and see if the transformer VA rating is sufficient enough to supply the demand.

This may seem unimportant to you at this time, yet there can be many hours of self-created problems for you if you don't remember these rules. For example, you may have a relay fail in a unit with holding coils that are at the limit of supplied VA. The replacement you use may have a slightly higher VA rating. After all of the wiring and mounting is done, you turn the power on and nothing happens except the holding coils begin humming. The new has caused the voltage of the secondary to drop far below the 24 volts required to operate the holding coils.

DISTANCE BETWEEN TRANSFORMER AND CONTROL

Another important factor to remember is the distance you have to string wire. In many instances, a technician may get carried away with how he is wiring a unit. Control wire is easy to run when you use 18/3, 18/5, or 18/7. Rule of thumb in most wiring is to increase wire size when the run of wiring exceeds 50 feet. This holds true to low-voltage control wiring. Excessive length of wire can cause a voltage drop due to the resistance of the wire. If you are wiring a small commercial unit across the roof of a shopping center and exceed the rule, that being, not to exceed 50 feet from transformer, when the power is turned on, in all probability the unit will not operate. In some cases, units will operate and then start gathering a history of burned holding coils. Improper voltages overheat the fine wires of the windings of the holding coils. Each time a technician will troubleshoot the system, he will probably replace the control and walk away from the unit. If it continues to happen to the same technician, he will eventually correct the problem. Don't make the mistake in the beginning and you won't have to worry about correcting it. Increase the wire size for every 50 feet of distance away from the transformer the control will be. Go from 18 to 16 to 14 and so on in gauging your wire sizes.

Sometimes a manufactured transformer will have a fuse on the secondary side; this helps avoid damage to the secondary windings. It is possible for a defective holding coil to destroy a transformer. If fusing is not supplied by the manufacturer, you can install one yourself, on the job-site. A fuse-holder and fuse can be purchased in most automobile supply and parts stores. Select one that is small enough. If you use a 10-amp fuse, the transformer secondary windings will melt before the link in the fuse. There are fuses and holders that work fine within the ranges of protection needed. A 40-VA transformer can be protected by a 1.5-amp fuse. Figure 6-2

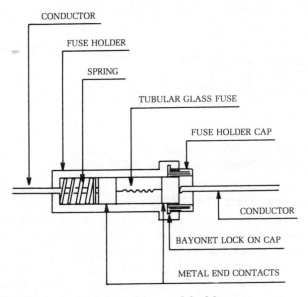

CONDUCTOR

FUSE HOLDER

SPRING

TUBULAR GLASS FUSE

FUSE HOLDER CAP

CONDUCTOR

BAYONET LOCK ON CAP

METAL END CONTACTS

Fig. 6-2. Automotive type fuse and holder.

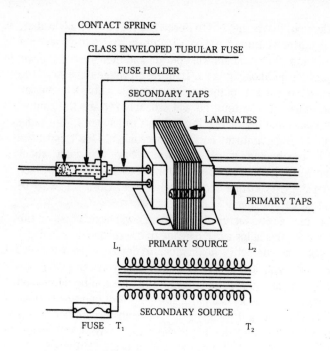

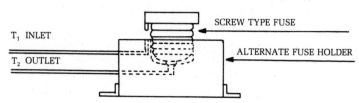

Fig. 6-3. Illustration and schematic of how to install protection fuse for a transformer.

shows the automotive type fuse and holder. In the schematic shown in Fig. 6-3, the placement of the fuse into the circuit is shown.

STEP-UP TRANSFORMERS

Manufacturers have placed on their data plates, electrical limits that their equipment will operate properly with through a range of voltages. For example; a plate might read 208/230 volts ac. If this unit is installed at a location that is being supplied 220 volts ac, it should operate without a problem in its electrical components for many years. The reason is that there is enough of a built-in buffer for electricity variations. Sometimes you will find that, at time of peak loads in certain geographical locations, full voltages are falling under the rating that they should be. A circuit that usually has a voltage of 115 volts ac might fall to 98 volts ac. Light bulbs are dimmer as a result of the low voltage. If you apply the same principle to a motor, which is turning a compressor, the motor will overheat since it is drawing more current in order to provide the same power at lower voltage. The motor windings might not reach the point of causing an overload to trip, so it continues operating at a high temperature causing the insulation on the motor windings to deteriorate and eventually fail.

For this reason, a transformer was designed to raise the voltage value above the supply voltage. This step-up transformer (called buck and boost in some areas), is quite large and heavy. In older communities, power companies supplied 200 volts ac. The 208/230 volt ac equipment is the most commonly sold. Special orders to the manufacturer can be made in some cases for a 200 or 208 volt unit. They are expensive and take a long time to get delivery, if it can be gotten at all. The only other option, instead of the continual replacement of motors, is to install a step-up transformer. The actual supply voltage has to be exactly matched to the primary of the transformer. The secondary voltage and amperage capabilities must be within the limits of the amounts specified on the manufacturer's data plate mounted on the equipment.

The 10% electrical variance rule of thumb should be remembered.

Example: 240 volt ac motor
       $\times$ .10 percent voltage variance
       24 volts

If the electric motor powering the compressor is rated for 240 volts, and you subtract the variance, you will find that this motor enters the destructive zone at 216 volts ac. The rule holds true for all ac-operated equipment.

TESTING TRANSFORMERS

Continuity testing is in order here for both windings. This is only done when the presence of voltage is confirmed. All too many times a technician assumes, this is not a good habit. Don't ever assume anything. There might come a time when you have to go back over a job done by another technician, don't assume what he has checked already. Start from the very beginning; you never know what kind of a day the other technician was having the day he checked the unit. Many transformers have been changed after being condemned. The unit still wouldn't operate. Further investigation shows an open in the primary circuit. The transformer must have the proper primary voltage in order to produce the proper secondary voltage. The problem wasn't the transformer and could have been repaired quickly with a splice or a wire nut.

CHAPTER 7

Circuit Breakers

The circuit breaker is an electrical device that opens an electrical circuit and stops the flow of current. In this chapter I am going to explain three types of circuit breakers that are used in our industry. Before getting into the text, let me inform you that circuit breaker manufacturers like to retain their individuality, making interchangeability of circuit breakers impossible. Whenever you have the need to replace a circuit breaker or a neighboring component part of the circuit breaker panel, take all of the information you can get from the panel. Manufacturers name, model, serial number, style (if any), and sometimes the dealer will ask for the color. The three types of breakers I want to explain are, the non-trip, quick-trip, and slo-trip.

NON-TRIP

In many parts of the country, electrical code provisions have been introduced to help prevent accidents. The requirement for a service disconnect within sight of any piece of electrical machinery is the best known of these provisions. With this type of protection, accidental electrocution and other mishaps can be avoided. This service disconnect code has been complied with in several ways. A simple lighting switch has been used on residential units. The double-pole, single-throw switch worked without any problems. A pull-type disconnect has been used. This is composed of a pull block that has

71

copper bus bars to make the circuit complete. When the block is pulled from the unit, the electrical circuit is opened. In Fig. 7-1 the typical pull-type disconnect is shown. A lever-type disconnect is also used. This type has bus blades mounted to a shaft. When the lever is pulled, the bus blades are pulled from the line lugs (receptacle where bus blades fit in to mate with the line side). Figure 7-2 shows this type of disconnect.

Fig. 7-1. Pull-type disconnect.

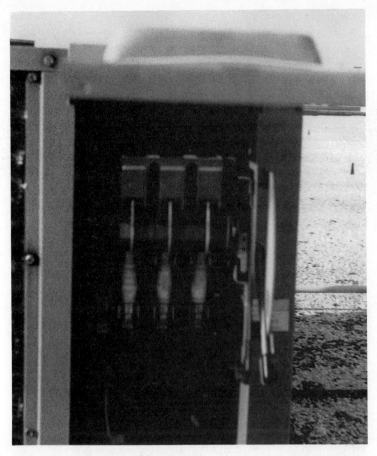

Fig. 7-2. Lever-type disconnect.

The non-trip service disconnect is an inexpensive and durable assembly. In Fig. 7-3 the disconnect can be seen as being very similar to an ordinary circuit breaker. Look very closely on the breaker assembly itself and the words "Non-Trip" will be stamped on it. There is an amperage rating for the unit also, perhaps this is where the confusion begins. The amperage rating applies to the amount of current the contacts in the switch will carry. Remember the amperage ratings that are used in relays and contactors; the same rule applies here. The higher the ampacity of this assembly, the more expensive it will be due to the contacts being constructed heavier to withstand the higher amperage loads crossing them. In many residential applications the 60-amp, non-trip breaker disconnect is

Fig. 7-3. Non-trip disconnect.

used. The main thing to remember is that this breaker will not trip if overloaded.

QUICK-TRIP

This type of circuit breaker is manufactured for light amperage loads such as household lighting circuits. They can be obtained in

120 and 240 volt ac ratings. Their ampacity range is very wide. This type of circuit breaker was designed to trip (open the circuit and stop the flow of current) when the rated ampacity is exceeded. One of the advantages of the circuit breaker is that you do not have to find a hardware store open in the middle of the night when a circuit breaker trips. The resetting device eliminates the need of having a box full of fuses that always seem to be the sizes you don't need.

TIME-DELAY BREAKER (SLOW TRIP)

This type of circuit breaker has been designed to conduct the locked rotor loads of starting motors and other initially high, resistive loads. These breakers are usually double breakers for 240 volts single-phase and triple breakers for three-phase equipment. Both types have a link that ties the circuits together mechanically. For instance in a three-phase compressor, the current in conductor L3 becomes excessive. The L3 switch begins to open, and as it moves, the mechanical link opens L1 and L2, thus eliminating the possibility of single-phasing.

Some equipment manufacturers use a circuit breaker such as these in their equipment as a service disconnect and to afford extra protection for the equipment. This will be found mostly on light commercial and heavy commercial equipment. The logic is that the compressor manufacturer will save money on warranty claims. In the installation the circuit breaker installed by the electrician might be too large, or more than one unit might be on the main breaker. This safety in the unit itself has proved to be a cost-saving device for many manufacturers.

TESTING

With a voltmeter, place one test probe on one of the screw lugs (screw that locks conductor into the circuit breaker) and the other test probe on the neutral bus bar. In the case of a 120-volt ac circuit breaker, 120 volts should be read on the meter. In the case of a 240-volt ac breaker, one probe should be placed on each screw lug of the breaker. The voltage should read 240 volts ac on the meter. If you test with one probe on the screw lug and the other on the neutral bus bar, each should read 115 volts ac.

Premature Tripping

This test is done only by overloading the circuit. With a snap-on type ammeter placed over the conductor of a circuit the amperage can be read. In a lighting circuit, the breaker rating is usually 15 amps. By placing an increasing load on a circuit such as lighting, lamps, toaster, or any other electrical device, you can slowly raise the amperage load on that conductor while observing the increasing current. If the circuit breaker is rated for 15 amps and trips when the load reaches 11 amps, the breaker should be replaced.

To check an electric motor circuit, the same principle applies. With a machine there is a slight risk of damage. If a unit is old and on its last leg, you might put the finishing touches to it by performing this test. When a compressor trips the breaker and there is a doubt whether the circuit breaker is defective or not, perform the following test. With the ammeter applied to the conductor that supplies the compressor, turn the compressor on and observe the amperage. It will probably be operating within the limitations of the circuit breaker. In order to cause the amperage to raise, cover the condenser fan inlet or outlet. This will cause the compressor to operate at higher compression pressures, thus causing a higher current to flow through the circuit breaker. Observe your ammeter to make sure the amperage doesn't exceed the FLA of the compressor.

Continuity

With the circuit breaker removed from the panel, a continuity test can be made to see if the switch resets and to measure the amount of resistance across the contacts. Figure 7-4 shows a typical circuit breaker entrance panel. The one in the picture has screws around the perimeter of the cover. Some have screws inside the cover that removes a cover plate. Removing these screws very carefully and removing the cover without touching any of the wiring will expose the conductors to the different circuits in the house. Notice there is a main circuit breaker; all of the breakers can be turned off including the main, this will decrease the chances of you grounding a circuit. Remember that there is still voltage in the panel. The only way to eliminate all the electricity from the panel is to remove the power company meter. This can only be done by an electrician or the power company. In Fig. 7-5 the two conductors from the electric meter feed the main circuit breaker. Figure 7-6 shows the two line bus bars that the circuit breaker engages with, to pick up current and transport it through the branch circuit conductor. Figure 7-7 shows the neutral bus bar which is the return conductor of the 115 volt ac circuit. The breaker can be removed

Fig. 7-4. Circuit breaker panel.

by disconnecting the conductor from its screw lug, then pulling up on the side that is in the center of the panel. The breaker plugs into the bus bar rail. Lift it out of the panel and now you can either test it or replace it. Figure 7-8 shows a typical circuit breaker.

In many locations, it is not a bad idea to check a circuit breaker panel from time to time to see if the wiring to the circuit breaker lugs is tight. In some cases, wires can work loose from vibrations

Fig. 7-5. Circuit breaker panel with cover removed. Two main feeder conductors are indicated.

and expansion; in the case of aluminum wiring this condition is especially true. Claims have been made that there is a distinct possibility that loose wiring in the entrance circuit breaker panel might cause an arcing that results in fire. On a service call, if the circumstances warrant it, it is a good idea to check the wire connections to each of the circuit breakers in the customer's panel. You and the customer will sleep better knowing that you checked a possible fire potential.

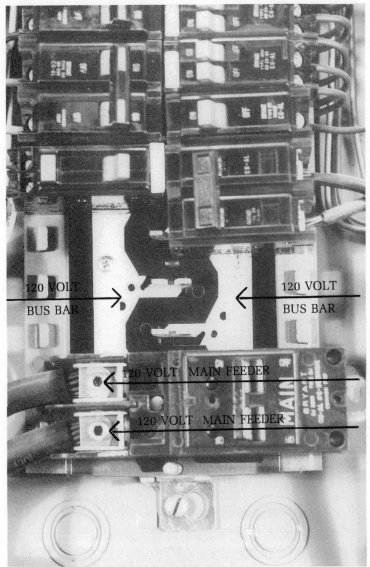

Fig. 7-6. Breakers removed to show the two bus bars providing 120 volts each thus giving 240-volt service in the structure.

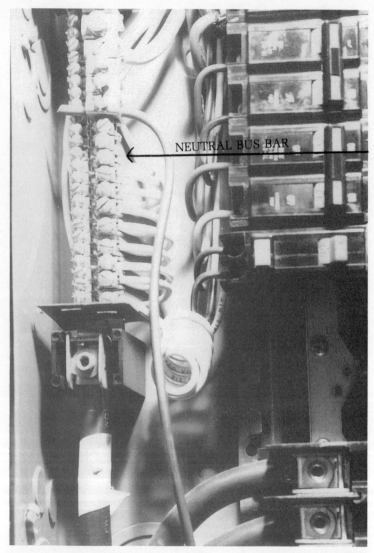

NEUTRAL BUS BAR

Fig. 7-7. Neutral bus bar used for one leg of all 120-volt circuits in the structure.

Fig. 7-8. Typical linked circuit breaker, with double thickness which straddles both bus bars, protecting 240-volt supply.

CHAPTER 8

Motor Starters

This chapter is strictly for those technicians that are servicing light commercial equipment. In most residential equipment, a contactor is used to start an electric motor, whether it is a compressor motor or a pump drive motor; however, the contactor offers no built-in protection for the motor it operates. The motor starters are designed to operate electric motors, but have an over-current protection by using overloads. The overload is similar to the circuit breaker in the way it acts upon the motor starter contact. Figure 8-1 shows a typical motor starter.

The control wire that causes the magnetic switch to close the contacts, is series-wired through bimetal auxiliary switches. As you have learned, bimetal will flex when heated or cooled. L1, L2, and L3 are wired from the line side through heaters to the load side. These heaters are designed with a specific resistance that causes them to produce heat when the designated limit is reached. The heaters can be designed for very close tolerances. They heat the bimetal very quickly and cause the control circuit to open. This in turn opens the contacts of the starter and the flow of current is stopped. The bimetal switch is a manual reset type. When the heater cools, the motor will not begin to run again until someone manually resets the switch. The advantage of this arrangement is two-fold. One, the tripping amperage is sized to the exact amount a motor can conduct before damage is done. Second, if a problem exists with the motor, it will not try to restart itself. A technician would check

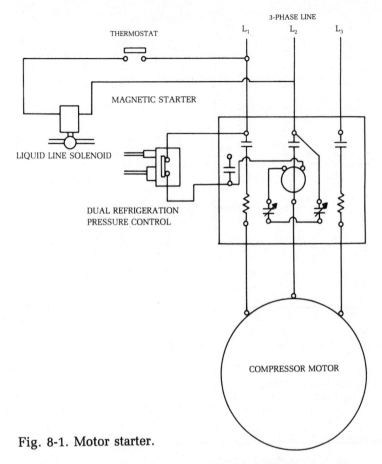

Fig. 8-1. Motor starter.

things before attempting to start the motor again. You should use an ammeter to confirm whether the heaters and bimetal switches are functioning properly. If they are, you might find a bad bearing, belt, or some other source causing the motor to drag and overheat.

This type of installation might be seen in a residence that uses a pool circulating pump, or perhaps a lawn sprinkling system with a pump large enough to be protected with a motor starter. This type of motor starter is considered to be automatic and can have a remote switch somewhere to turn it on and off. There are manual motor starters that have the same overload protection as the automatic ones, but they must be turned on and off manually at the switch. Figure 8-2 shows this manual type of starter. These switches have replaceable heaters that can be sized to any amperage load, whether it is a single-phase or three-phase application.

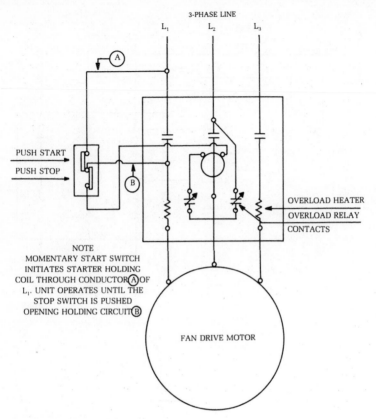

Fig. 8-2. Manual motor starter.

AUXILIARY SWITCHES

An auxiliary switch or switches will be found mounted on some of the automatic motor starters. These are usually n.o. switches, that close causing operation of a fan or other appliance or appendage in the system. If a water pump is located in a small room or closet, an exhaust or intake fan might be energized whenever the pump starter starts the pump. The n.o. auxiliary contacts close turning on the fan. Some starters can be acquired with more than one auxiliary set of contacts.

This is a fitting time to go into a little more detail about the wiring of three-phase motors. Three-phase motors have six windings. Wire leads from these windings are usually brought into a terminal box by the manufacturer. It is here in the terminal box that the branch circuit wiring is attached to the motor windings. Usually there are nine leads brought from the windings; they are marked with bands showing their numbers. Each wire lead having a different number from one through nine. In Fig. 8-3 you can see a wiring schematic of a typical electric motor circuit. Motor voltages vary with the location and application for which they are manufactured. Let me state again, this book is not a pertinent text for heavy commercial and industrial motors. Their voltages are different from residential and light commercial application; for this reason 120, 240, and 480 volts are discussed. If a motor is rated for two voltages, it will be up to you to wire it for the correct supply voltage. Make sure of the voltage; it doesn't take long for a winding to burn, when it is designed for 120 volts and 240 volts is put through it.

In our industry, L1, L2, and L3 designate the line sides of a device. The line side is the conductor that has the voltage constantly ready to flow. The other side, designated by T1, T2, and T3, is

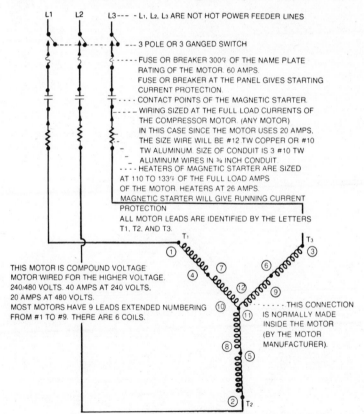

Fig. 8-3. Wiring schematic of the average motor circuit.

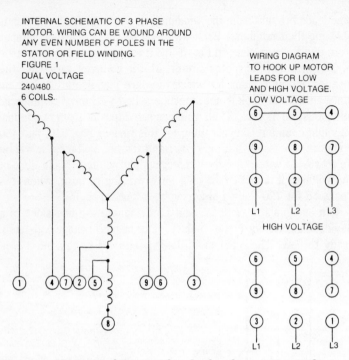

INTERNAL SCHEMATIC OF 3 PHASE
MOTOR. WIRING CAN BE WOUND AROUND
ANY EVEN NUMBER OF POLES IN THE
STATOR OR FIELD WINDING.
FIGURE 1
DUAL VOLTAGE
240/480
6 COILS.

WIRING DIAGRAM
TO HOOK UP MOTOR
LEADS FOR LOW
AND HIGH VOLTAGE.
LOW VOLTAGE

HIGH VOLTAGE

Fig. 8-4. Wiring schematic for dual-voltage six windings.

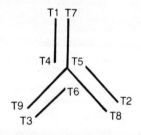

VOLTAGE	L1	L2	L3	TIE TOGETHER
LOW	(T1,T7)	(T2,T8)	(T3,T9)	(T4,T5,T6)
HIGH	T1	T2	T3	(T4,T7) (T5,T8) (T6,T9)

Fig. 8-5. Wiring schematic for a single-speed, delta-connected, dual-voltage electric motor.

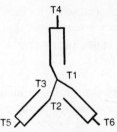

SPEED	L1 L2 L3	INSULATE SEPARATELY	TIE TOGETHER
LOW HIGH	T1 T2 T3 T6 T4 T5	T4-T5-T6 . . .	. . . (T1,T2,T3)

Fig. 8-6. Wiring schematic for a two-speed, single-winding, variable-torque electric motor.

the load side of the device. The load side is the conductor to which the device is wired. The T side doesn't carry any voltage when the control switch is in the open position. Figure 8-4 shows the hookup of the windings and line side for dual voltages. This motor has a low voltage of 240 volts ac. Figures 8-5 through 8-7 show different wiring schematics with Y windings along with dual speed motors. This will be an easy reference on a job someday.

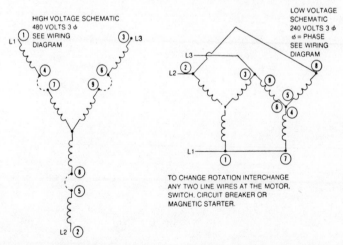

Fig. 8-7. Wiring schematic for a single-speed, Y-connected, dual-voltage electric motor.

Don't forget the importance of rotation. It is very important that you remember this minor point. Fan blades, water pumps, some compressors, just to mention a few need to have their drive motors turning in the proper direction. This holds true for single- and three-phase motors. Some single-phase motors do not have reversing capabilities. These are fixed rotation motors. Reversible rotation single-phase motors accomplish this by electrically reversing two connections in the motor. As I stated in the beginning of the book, three-phase motors are easily reversed by reversing two lines. By reversing L1 and L2 with each other, the rotation of the motor will reverse. This procedure was discussed in Chapter 3 in the section about compressors with locked rotor.

CHAPTER 9

Safety Controls

Safety controls are designed for the protection of the component parts of a refrigeration or air conditioning system. The basic controls such as, fuses, circuit breakers, and overload relays were covered in earlier chapters.

The amount of protection a system receives depends upon the purchase price of the unit. In order to remain in a competitive market, many manufacturers have a basic stripped model without safety controls. The more protection a unit has, the more expensive the retail price. Many of these units can be seen in multiple family structures throughout the country. The purchaser is very price conscious when signing a contract buying a home. As long as the air conditioning cools, they are happy. Unfortunately, this equipment is in the out-of-sight, out-of-mind category.

In this chapter, I'll cover some of the safety controls that you will encounter on some of the units you service, or you might want to install a safety device on a unit that doesn't have one. Regardless of how complex a control looks to you, remember, there are three variables that can cause it to operate. Pressure, temperature, and humidity, or a combination of these are used to activate a safety control.

I am placing these safety controls into two categories: those that protect the system electrical components, and those that protect the system mechanical components. Only those used on residential and light commercial units will be covered.

Fig. 9-1. Overload relay, clamp-on type.

ELECTRICAL SAFETY CONTROLS

Klixon Overload Relay

Figure 9-1 shows the typical overload relay used for the protection of the windings in the compressor. This type is fastened to the compressor body in a way that heat is transferred from the steel body to the overload sensor. In this way, the Klixon can sense excess heat in the steel compressor shell. There are heaters within the Klixon that generate heat when the current drawn by the compressor reaches a pre-determined amount. In this way, the Klixon overload relay will interrupt the current flow if either condition is sensed. There are slight lag times built into these relays. This eliminates many pre-mature trips upon start up. There are specific sizes to be used with specific compressors. In general, the control will carry an overload for about six to 16 seconds in those that are rated under 100-amp service protection. Relays that service loads over 100 amps would have a trip time of about two to six seconds. This is just another rule-of-thumb to give you an idea of the type of protection that can be expected from this type of control.

Overload Relay (Remote Mounted)

There is a big difference between this overload relay and the one mounted to the compressor. This type would be found more commonly on the larger tonnage units. I will describe one here in the event you encounter one installed on a smaller unit. Figure 9-2 shows a typical remote overload relay. Notice the size is comparatively larger than the Klixon. This device in most cases can be acquired with either a manual reset or an automatic reset. This kind of overload is called the hydraulic-magnetic type while the Klixon is called the disc type.

Time Delay Relays

The time delay relay has become a very popular safety device. It is a control that is directly responsible for extending the life of the compressor. Its function is to keep the compressor from trying to start until a certain time has lapsed from the time it stopped operating. This gives the refrigerant within the sealed system time to equalize. By doing so, the pistons do not have excessive pressures against them when the compressor restarts. An electric motor trying to start with both a heavy normal load plus high head pressures placed on the pistons causes excessive heat in the wire windings, that eventually result in the insulation breaking down and the windings grounding out, or opening, or cross phasing. The time delay lets the pressure equalize avoiding this problem.

There are many examples that can be cited. Those people that never seem to get the thermostat set just right. They are constantly

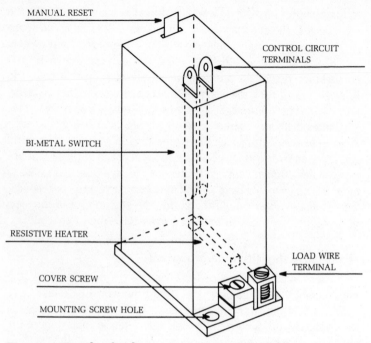

MANUAL RESET

CONTROL CIRCUIT
TERMINALS

BI-METAL SWITCH

RESISTIVE HEATER

LOAD WIRE
TERMINAL

COVER SCREW

MOUNTING SCREW HOLE

Fig. 9-2. Overload relay, remote (simplified view).

trying to make it hotter or cooler. All that they are accomplishing
is the murder of their compressor. A time delay prevents this. In
areas where brown-outs are common, that is when the power
company cuts power from one part of the grid to supply another,
it is a good idea to install time delay relays. An instant power outage
can occur during the switching procedure and cause the compressor
almost to stop and then to try to restart. A time delay provides useful
protection to the unit in these situations. In areas where there are
extremely severe thunderstorms and the lightning strikes cause
frequent, short outages, it is also desirable to use time delay
protection.

There are solid-state electronic time delays that range in price
from inexpensive to expensive, depending what the application is
to be. There are mechanical types that include an electric clock that
interrupts the control circuit. The clock is coupled to switches that
accomplish this. After the designed amount of time is passed, the
clock activates a switch that completes the control circuit and allows
the compressor to start. In many instances, when time delay relays
are installed, it is a favorable condition to wire them into the circuit

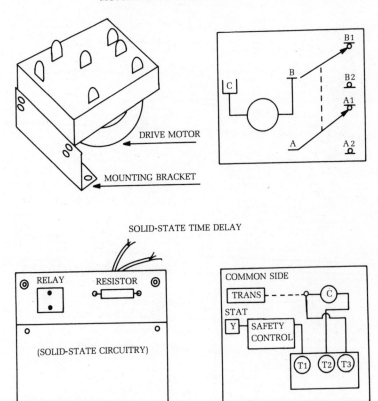

Fig. 9-3. Time delays, electronic and mechanical.

allowing the condenser fan motor to operate while the compressor is being timed out. The reason is, air movement in the compressor area and across the condenser usually helps to cool the refrigerant, which in turn drops the pressure. Figure 9-3 shows both types of time delays. The electronic device is easier to install and takes less time.

MECHANICAL PROTECTION CONTROLS

High Pressure Control

In Fig. 9-4 the typical high pressure controls are shown. One of these controls is primarily used in residential units while the other is used in commercial units as well as some refrigeration

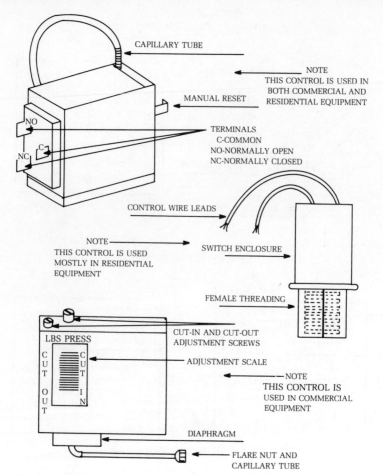

Fig. 9-4. High pressure controls, commercial and residential.

(commercial) equipment. The main difference is that one is factory adjusted, while the other may be field adjusted for whatever application it is being used for.

The control was designed to open the control circuit when a preset pressure is reached. This condition of high pressure exceeding factory specifications can be due to inoperative condenser fan motor, defective condenser fan blade, restriction in the discharge line of the compressor, or just an overgrowth of vegetation due to owner neglect. Regardless of the cause, the control opens and stops the compressor from becoming damaged. This type of switch can have a capillary tube that directs the high pressure to a remote switch, or it can be located directly in the discharge line as a diaphragm switch.

Pressure Relief Device

Due to Murphy's Law (if something can go wrong and break down, it will), a back up is needed in the event the high-pressure control malfunctions. A relief valve can be used in commercial units but isn't found often in light commercial or residential units. Figure 9-5 shows the relief valve. This device is costly and would raise the price of a home system. For this reason, other pressure relief devices were introduced. In Figure 9-6 a typical relief plug is shown. The center of the plug has a hole machined into it. This hole is then filled with solder that melts at the temperature reached when the refrigerant pressure exceeds the specified limits. When the solder melts, the refrigerant is released from the system. In some units, one of the factory soldered joints is made with soft solder. All of

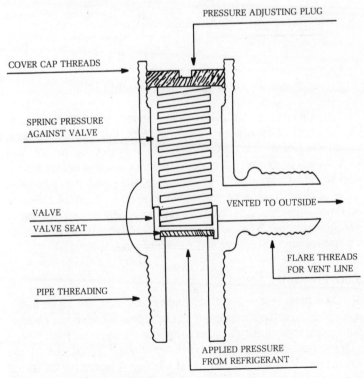

Fig. 9-5. Pressure relief valve.

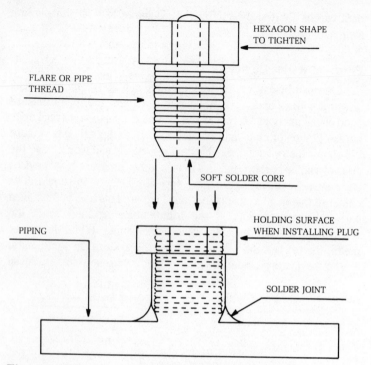

Fig. 9-6. Pressure relief plug, with fusible plug.

the other joints are hard soldered with some type of silver content. In the event of excessively high pressures, the joint will rupture, open, and release the refrigerant from the system before anything is damaged. In commercial systems, where the amount of refrigerant lost would be quite expensive to replace, the costly pressure relief valve is used. This valve only opens long enough for the pressure to drop within specified limits. In this way, only a small amount of refrigerant is lost. The cost of the valve is offset in the savings made by not having to recharge the complete system.

Low Pressure Control

You must remember that ample refrigerant has to be in the system so that the compressor has proper lubrication and cooling. The refrigerant is the oil carrier through the system. Each time a compressor starts, it pumps a large amount of oil out of the crankcase. It usually takes several minutes of operation for the oil to return to the crankcase. Also the suction gas returning from the evaporator cools the motor windings of the compressor. So what happens to a compressor that operates without refrigerant? You're right, it will destroy itself. It is for this reason that a low pressure control is used on some systems. A preset pressure opens a switch when it is reached. Again there are both factory preset and adjustable types. Many of these open when a low pressure about 15 pounds is reached. This control, like the high pressure control, can have a capillary tube or a direct mount diaphragm. The high and low pressure controls look alike. One way to tell them apart is by the size of their diaphragm. The high pressure control has a small one and the low pressure control has a large one. Plus the scale of pressure will be different. In commercial applications, a dual-pressure control can be used. This is a combination of both controls enclosed in a common housing. The added advantage to this control is that it will stop the machine when it still has a positive pressure within the refrigeration system. This prevents air from entering the system and introducing moisture to it. The service technician is usually called before atmospheric pressure is reached inside the system.

Compressor Ambient Control

This is a small bimetal limit switch that opens on temperature fall. It can be located in the outdoor condensing unit. This switch can be wired to prevent the compressor from operating on those days when the outside temperatures are low.

Supplemental Heat Control

This bimetal switch is wired to electric strip heaters to prevent them from energizing on temperate days. Both of these devices became popular during the energy crisis. I'm telling you about them if you choose to use them, or if you should come in contact with a system that uses these devices.

Fan Cycling Control

This control can be one of the most important in some geographical areas, especially the cooler latitudes. In order for an air conditioning system to operate properly, the condensing temperatures are operating within certain limits. If the limit is exceeded on either side of the scale, too high or too low, the efficiency of the unit is lowered. For his reason, the temperature of the condenser can be controlled by cycling the condenser fan motor. Some commercial units of higher tonnage may also use a damper device to prevent a cold prevailing wind from blowing across the condenser coil causing the head pressure to remain too low. The damper is actuated by the rise and fall of the head pressure. In some residential units, a two-speed condenser fan motor is used to attain better control. The fan-cycling is accomplished with one control. The fan is either off, low speed, or high speed.

I've been asked, ''why don't they just open the windows?'' In certain circumstances, perhaps they would, however in many cases it would be impractical. This type of control holds true for low ambient operation of refrigeration equipment such as walk-in-freezers and ice making equipment.

De-Ice or Defrost Control

This control is found on heat pumps. The device has either a temperature or pressure-actuated clock that places the condensing unit of a heat pump into a defrost cycle when the ice buildup on the condenser inhibits its performance. The cycle is terminated with the same control when a high limit is reached. This control has a switch on it that also might energize the auxiliary heat strips located in the electric furnace of the evaporator section. This is discussed in detail in the chapter covering heat pumps.

The air conditioning manufacturers usually have several models to be price competitive throughout the world. Depending on the model you own or are repairing the number and quality of the control devices will vary. Many manufacture some of their own controls, but for the most part, there are a few companies that manufacture controls for the different air conditioning units. I've found that in certain areas, some brands of controls are more dominant than others. Remember, they all do the same thing, basically. There are many other controls on the market today, and if you look over a control catalogue, you can learn how to add safety devices to a system that has none.

CHAPTER 10

Evaporator Section (Air Handler)

Depending on your location geographically, the evaporator sections might differ. In northern latitudes, heating fuels in most cases are either oil or gas. Most fossil fueled furnaces are equipped with an A coil evaporator, or something similar. It is usually mounted in a plenum above the furnace. The fan section is located below and usually consists of a direct-drive or belt-driven, squirrel cage fan.

In the temperate and southern latitudes, the electric furnace along with a heat pump is very efficient. In many of these units, the transformer is located in the evaporator section. Along with the transformer, there are safety controls.

HEAT STRIPS

Limit Switch

In Fig. 10-1 an electric strip heater assembly is shown. The limit switch is found in the electric furnace as it is in the fossil fuel furnaces. Its main purpose is to protect the unit from reaching excessively high temperatures. The electric furnace can overheat if there is a fan failure or if not enough air is being blown across the resistive elements. Figure 10-2 shows a typical limit switch. Some of these are dual-purpose switches. The first stage of heat rise is wired to control the fan. In this way a blast of cold air is not felt when the electric furnace begins to operate. When the air in the plenum

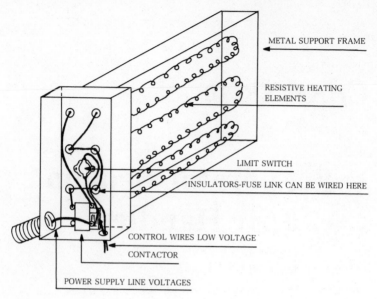

Fig. 10-1. Heat strip assembly.

reaches a comfortable, heating temperature, the fan will start operating. If the heat continues to climb, a second set of contacts contained in the limit switch opens. Figure 10-2 shows a surface mount limit switch. On some applications an insertion type, as shown in Fig. 10-3, is used.

Fuses

In the event a limit switch should fail, there is a back-up protection called fusible links. Unlike the regular fuse you have worked with, the links are constructed somewhat differently. In Fig. 10-4 two of the typical fusible links are shown. When these fuses open, they must be replaced as an assembly. They only open if the heat or amperage exceeds the unit limits.

Sail Switches

This switch is seen primarily in commercial units, and in my opinion it is a valuable safety switch for both heating and cooling. It is generally located downstream of the evaporator fan. The common of the low-voltage control circuit is wired through the sail switch. The Y terminal (cooling and generally wired to the compressor) or the W terminal (heating and used with electric

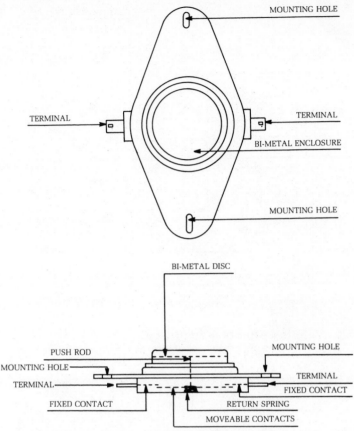

Fig. 10-2. Limit switch surface mount.

furnaces) is wired in series through the sail switch. In Fig. 10-5 a sail switch is shown. There has to be a certain volume of air blowing across the sail switch for it to close its circuit. Diaphragm type of pressure switches can be used also. These are found in some light commercial applications. With this control, if filters are dirty, or a fan belt is broken, or the fan motor isn't operating, the system will not operate preventing possible damage to any of its components. In units not equipped with limit switches, this type of switch is invaluable.

Sequencers

These are shown in Fig. 10-6 and are commonly used in electric furnaces to stage heat. In most residential units, resistive heat strips

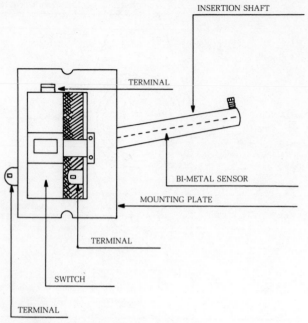

Fig. 10-3. Insertion type limit switch.

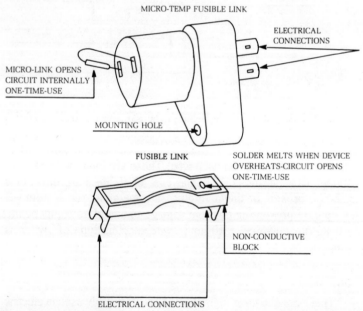

Fig. 10-4. Fuse links.

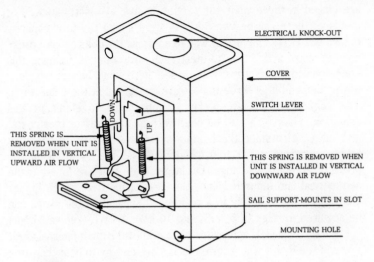

ELECTRICAL KNOCK-OUT

COVER

SWITCH LEVER

DOWN

UP

THIS SPRING IS REMOVED WHEN UNIT IS INSTALLED IN VERTICAL UPWARD AIR FLOW

THIS SPRING IS REMOVED WHEN UNIT IS INSTALLED IN VERTICAL DOWNWARD AIR FLOW

SAIL SUPPORT-MOUNTS IN SLOT

MOUNTING HOLE

NOTE
WHEN INSTALLED IN A HORIZONTAL AIR FLOW
BOTH SPRINGS ARE REMOVED

Fig. 10-5. Sail switch.

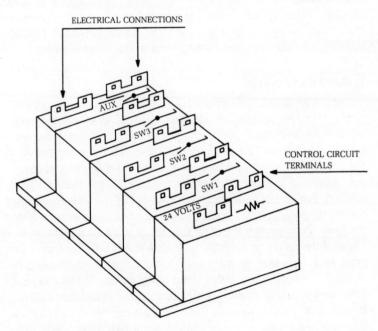

ELECTRICAL CONNECTIONS

AUX

SW3

SW2

CONTROL CIRCUIT TERMINALS

SW1

24 VOLTS

Fig. 10-6. Sequencer.

are placed in the evaporator section in five kW increments. For instance, there can be two 5-kW heaters giving the total capacity of heat of 10 kW. The proper way to energize the units and not cause a severe power drain, is to start each with a delay. Imagine this operation with three strips, you can see there would be a tremendous power demand if all three units were energized at the same time. The sequencer, or time delay switch, helps to eliminate this condition. These can also be used with an appropriate thermostat and sub-base to stage the heating system. All strips do not have to be energized if the thermal demand isn't present. This too provides an energy savings. On larger units that are found in the commercial and industrial line, the heat strips have much higher kilowatt ratings. Usually in a residential application the strips can be acquired in sizes of 5, 7.5, and 10 kW. A kWh (kilowatt-hour) of electricity will yield approximately 3,400 Btu (British thermal unit). You can see that it will require much more energy to heat the same size area, that is being cooled, with an electric furnace.

The sequential switch operates with bimetal switches. The first switch to close is actuated by the thermostat within the conditioned space. The first bimetal switch will flex causing a current flow in the second switch which will heat the bimetal and cause it to flex causing the third switch to start heating the bimetal, and so on. When the control voltage is removed, the bimetal switches begin cooling and drop out the resistive heating coils in the reverse order.

Water Float Switch

In certain installations where the evaporator section was installed in the attic, an auxiliary condensate drain is used. If the main condensate drain pan should become restricted, the condensate water will overflow into the auxiliary drain pan. A float switch that has the control circuit wired through it is mounted inside this auxiliary pan. When the float rises, the switch opens and stops the unit. This auxiliary pan should be separately piped out of the attic to a place outside the structure that would be obvious when water dripped from the drain. Perhaps near the front entrance would be a good location. There if the owner of the structure sees water dripping from that pipe, he knows that his main drain line for the air conditioning is clogged. This procedure helps avoid costly ceiling repairs and perhaps damage to the furniture located beneath the evaporator section. It helps if the service technician installing a water float switch leaves a note in the condensing unit stating that a float switch is in use.

The same note can leave instructions as to circuit breaker numbers or any other data that would be helpful.

Run capacitors are usually located for the evaporator fan motor inside the cabinet of the evaporator section. A fan relay is also located in the same section.

CHAPTER 11

Thermostats

Many technicians, including myself, consider the compressor as the heart of the air conditioning system. It pumps the refrigerant through the system as the heart pumps blood through the body. If the compressor is the heart, then undoubtedly the thermostat must be considered the brain of a system. It tells the system to heat, cool, or ventilate, and exactly how much of that to do.

There are many types of thermostats on the market today. Some of them are very simple and others look like something from another planet. Price determines their intricacy, and they can have microprocessor circuitry. These mini-computers can be programmed to adjust or change temperatures automatically during the week. They can be programmed to shut the system down if the structure is to be closed and vacant for a while. They turn themselves on automatically prior to the occupants returning in order to pre-cool the structure. These along with other advantages can be found in the new technology being brought into the thermostat manufacturing.

In Fig. 11-1 the popular round one is, in my opinion, number one of the workhorses of thermostats. It is a bimetal control that uses a mercury switch. When the bimetal flexes in either direction, the mercury switch closes in the way the mercury tips. In this way, both heating and cooling can be accomplished through the same mercury switch. With the addition of multiple mercury switches, staging of heating and cooling is made possible. The thermostat shown in Fig. 11-1 is only a switch. This switch mounts to a sub-

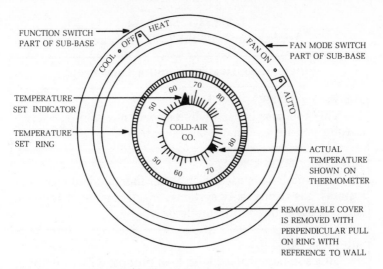

FUNCTION SWITCH
PART OF SUB-BASE

HEAT

OFF

COOL

TEMPERATURE
SET INDICATOR

TEMPERATURE
SET RING

60 70

50 80

COLD-AIR
CO.

80

50

60 70

FAN ON

FAN MODE SWITCH
PART OF SUB-BASE

AUTO

ACTUAL
TEMPERATURE
SHOWN ON
THERMOMETER

REMOVEABLE COVER
IS REMOVED WITH
PERPENDICULAR PULL
ON RING WITH
REFERENCE TO WALL

Fig. 11-1. Honeywell T87F thermostat.

base. It is the sub-base that contains the specific control circuitry. Table 11-1 lists some of the more popular round sub-bases giving part number and function. Honeywell is about the largest manufacturer of thermostats. They manufacture them for other companies and place the other companies names on the face of the thermostat, in essence they are Honeywell products. There are other manufacturers however producing their own products that work along the same line as do the Honeywell products.

In Fig. 11-2 a sub-base is shown for a round thermostat. Note the indexing lines on the sub-base. It is very important to plumb this type of thermostat when installing. Vibration of the wall can also change this adjustment. Ordinary use with a customer turning it on and off swinging the levers can throw the sub-base out of alignment. If this happens, the calibration of the thermostat is wrong. This means that the desired temperature settings will not be accomplished, due to mis-alignment to the sub-base. A small wrench tied to a string makes a good plumb-bob.

You can see that the mounting surface has to be stable in order to use a mercury-switch type thermostat. Any vibration or quiver in the wall can cause short-cycling of the condensing unit. A door slam shouldn't cause a vibration that would cause a mercury switch to act, but there are many cases where this does happen. Some are found in mobile homes, modular homes, and in certain cases where

Table 11-1. Sub-base
Applications with the T87F Thermostat.

Single Stage Heating/Cooling Thermostat
Sub-Base: Q539C   This sub-base is used when system has cooling only. It is equipped for automatic fan operation.
Sub-Base: Q539A   This sub-base is used with either independent heat-cool systems, gas electric split systems, or self-contained systems. Fan operation in cooling mode only.
Sub-Base: Q539J   This sub-base is used with heat pumps, with or without accessory electric heat strips. Automatic fan operation is available with heating or cooling.
These are but a few of the sub-bases used with the T87F thermostat. This table illustrates the versatility of the one specific thermostat. This also applies to the rectangular models. The manufacturer can supply a catalogue that will show the applications for each specific product.

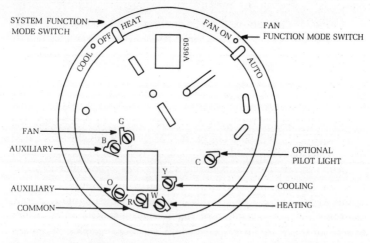

Fig. 11-2. Sub-base.

Fig. 11-3. Bi-metal thermostat.

aluminum studding for the wall has been used. Be aware that there are other options to you for a thermostat selection.

Figure 11-3 shows a bimetal thermostat with a magnetic assist. This type of thermostat would be more appropriate to an area with slight vibration.

Another type of thermostat that can be used in this situation is the bellows type. It is a gas-filled bellows that expands and contracts with heating and cooling. This movement actuates the switching device. Many of these thermostats are manufactured to use line voltage.

If you have a situation where there is a vibration in a wall that is causing erratic thermostat operation, it might be more cost effective to change the thermostat style than to pull wiring through a wall in order to change the thermostat location. Thermostats are built in different shapes. They can be round, square, or rectangular, but remember that they all perform the same job.

The last one that can be used is the electronic thermostat. It uses electronic switches with thermistors. The basic theory is that no energy flows when resistance maintains equal potential in the circuit. An imbalance caused by the heating or cooling of a thermistor changes resistive value and allows a flow of current. Of course the cost of this unit is very high. In most cases, installation also takes time due to the fact that additional wiring is usually required.

In my opinion the thermostat and the expansion valves are the most unjustly accused components of the system. They are both very dependable and are sometimes falsely condemned. Remember the thermostat is nothing more than a series of switches that open and close telling the system what to do.

WIRING

The thermostat is the center of operation. It receives wires from the evaporator section and from the condenser section. Number 18 wire is usually used. It can be 18/2, which designates 18-gauge wire, two conductors. There is a difference in price between 18/2 and 18/7; however, in the event of a wire breaking or shorting, it is a lot cheaper and faster for you to have a couple of extra conductors pulled through the structure. In fact there may come a time that another control is to be added to the condensing unit or evaporator unit, the wire will be available to carry another circuit. This is the reason I suggest the use of 18/7 thermostat wire. Each of the conductors insulation has a different color. This makes the mating of different circuits easier.

Reviewing Chapter 5, remember that the relay has a common wire to it. In the case of a 24-volt control circuit, each relay has T_1 from the transformer wired to it. This means that each relay's holding coil only needs the T_2 of the transformer to energize its electromagnet coils.

SUB-BASE TERMINAL IDENTIFICATION

R. This is the terminal that accommodates the T_2 wire from the thermostat. When this voltage passes through the thermostat switch, a relay/relays will energize. In certain applications where there might be two separate control circuits, such as those found in some fossil fuel units, a separate transformer is used in each system. One would be for the cooling that would route through the R_1 terminal, and one perhaps a different voltage through R_2. If both heating and cooling controls are of the same voltage, a jumper wire can be placed between R_1 and R_2. Some thermostat manufacturers use V (voltage) or RC in place of the R. It is the same terminal and the common from the transformer is affixed to it.

Y. This is the terminal that is used to energize the condensing unit, of course in the cooling mode. Depending upon the application, there might be a few stages to cooling, thus Y_1, Y_2, Y_3 terminals indicating first, second, and third stage of cooling.

W. This terminal is used to energize the heating. As with the cooling, the heating mode might have more than one stage, thus, W_1, W_2, W_3, would be used. This would indicate three stages of heating.

G. This terminal is used for the evaporator fan. In some cases it might be indicated with an F.

C. This is a terminal where T_1 of the transformer is attached. The reason is usually for illumination of a pilot light. This could show when a compressor is in operation, or when a filter is clogged, or any other type of 24 volt signaling device used. In some thermostats, the C is used for cooling if there isn't a Y terminal.

O. This is an auxiliary terminal that will energize when in the cooling mode.

B. This is an auxiliary terminal that will energize when in the heating mode.

In Fig. 11-2, a round sub-base is shown with the terminal identification letters. Although the coding might be different depending on the manufacturer, the functions are the same. One more thing, after the wire is hooked to the sub-base plug, the hole where the wire comes through the wall should be sealed in order to prevent erratic thermostat operation due to hot or cold drafts that might be flowing in the wall. A cotton ball or something similar can be used.

DIFFERENTIAL

This word when used in relation to the thermostat refers to the span of degrees of temperature between the off cycle and the operating cycle. The ultimate differential is two degrees. In the mechanical type of thermostats, this is sometimes difficult to achieve. If a thermostat is set to maintain 76 degrees F., in a conditioned space, it would have to cycle on at 77 degrees F. and cycle off at 75 degrees F. thus giving you a two-degree differential. This can be attained with electronic thermostats but is quite difficult to achieve with the mechanical types. Some thermostats might have a very wide differential, causing the conditioned space to become too hot or too cold before it cycles on. Many thermostats can be calibrated, some have a fixed differential. Try to adjust to the two degree optimum.

LEVELING SUB-BASE FOR
MERCURY SWITCH THERMOSTATS

In Fig. 11-4, the plumb-bob is being used to level a round sub-base. Notice the alignment marks used. In Fig. 11-5 the rectangular

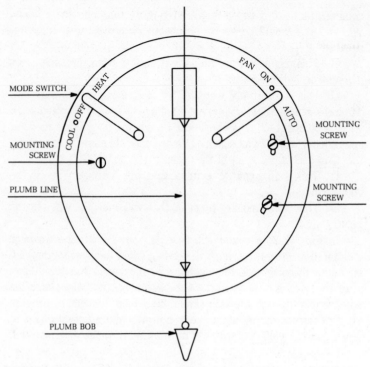

Fig. 11-4. Leveling sub-base with plumb bob.

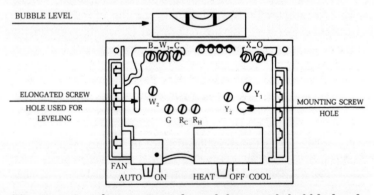

Fig. 11-5. Leveling rectangular sub-base with bubble-level.

sub-base is shown. This and the square type use a bubble-level to accomplish the leveling. There are two important things I want to mention at this point. First, be careful of marking up the customer's wall around the thermostat. Grease or dirt from your hands can mar the wall. Second, with some dry-wall construction, it can be difficult to get a secure mount for the sub-base. The result is a larger hole than the round thermostat will cover. A piece called a designer, or decorator plate can be purchased at the parts supplier. This commonly called goof, or oops or idiot, plate is nothing more than a larger circular piece of plastic that can be used to fasten to a different section of the wall where the screw holes have never been, then attach the sub-base to this plate.

PNEUMATIC

Don't let this word scare you. I'm going to touch lightly upon this type of controlling for those technicians that might come in contact with it. Up till now, the only means that were used for control were electrical, in the form of a switching device, or a bellows type that causes pressure exertion with expansion of the gas. The medium used in pneumatic controls is compressed air. Why is it used? Good question. Remember that for every 50 feet of wire run, the resistance factor dictates that the wire size be increased.

Where the control circuitry must be lengthy, compressed air lines can be more economical to install and maintain. A common air compressor is used to drive the entire system. There could be many units in an office building that will be controlled with the air pressure from the common air compressor. The air is transported through the building using either plastic or copper tubing as a conveyance. The air pressure finally operates pneumatic actuators that open and close dampers, or open and close switches. The switch is activated by air pressure. The PE (air pressure operates an electrical switch) switch can operate anything electrical as does the standard electromagnetic switch. The PE stands for pneumatic electric. The reverse of this is the EP switch. This EP (electric pneumatic) performs the exact opposite action. An electric switch controls the air flow. Those are the two main types of switches found in the pneumatics. Figure 11-6 shows a typical air compressor assembly used in the pneumatic control system. These systems are manufactured by Honeywell, Johnson, Robertshaw, and Barber-Colman to name some. The pneumatic thermostat as shown in Fig. 11-7 still uses the bimetal principle to bleed air from the source. By bleeding the air, pressure is not sent into the vessel that leads to either the pneumatic motor or the PE switch.

In Fig. 11-8 a pneumatic actuator is shown. This device converts the air pressure by the use of a piston to a mechanical force in the

113

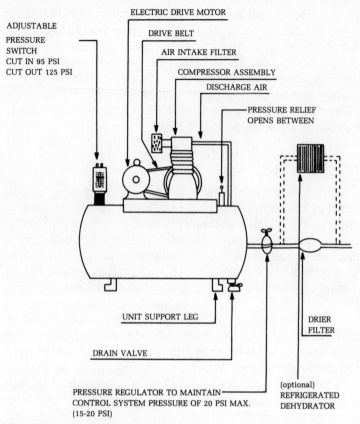

ADJUSTABLE
PRESSURE
SWITCH
CUT IN 95 PSI
CUT OUT 125 PSI

ELECTRIC DRIVE MOTOR

DRIVE BELT

AIR INTAKE FILTER

COMPRESSOR ASSEMBLY

DISCHARGE AIR

PRESSURE RELIEF
OPENS BETWEEN

UNIT SUPPORT LEG

DRAIN VALVE

DRIER
FILTER

PRESSURE REGULATOR TO MAINTAIN
CONTROL SYSTEM PRESSURE OF 20 PSI MAX.
(15-20 PSI)

(optional)
REFRIGERATED
DEHYDRATOR

Fig. 11-6. Air compressor system to supply air pressure in a pneumatic control circuit.

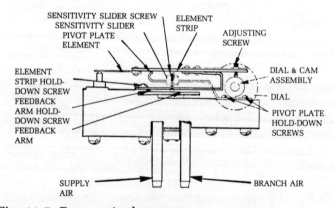

SENSITIVITY SLIDER SCREW
SENSITIVITY SLIDER
PIVOT PLATE
ELEMENT

ELEMENT
STRIP

ADJUSTING
SCREW

ELEMENT
STRIP HOLD-
DOWN SCREW
FEEDBACK
ARM HOLD-
DOWN SCREW
FEEDBACK
ARM

DIAL & CAM
ASSEMBLY

DIAL

PIVOT PLATE
HOLD-DOWN
SCREWS

SUPPLY
AIR

BRANCH AIR

Fig. 11-7. Pneumatic thermostat.

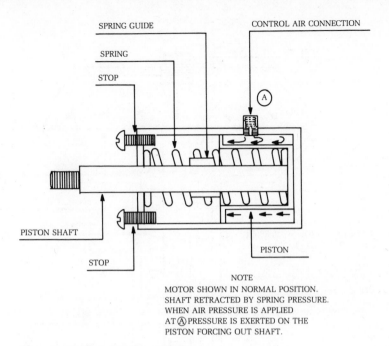

Fig. 11-8. Pneumatic motor.

push rod. This actuator can operate a valve, dampers, etc. Figure 11-9 illustrates a valve with a pneumatic actuator. Valves such as these are also made to operate with electric motors.

Alright, the next time someone mentions pneumatic controls to you, this basic knowledge will be helpful to you in understanding . . . if you didn't up till now.

ACCUSTATS

This is another type of thermostat that you may encounter. It is basically used in a commercial application where too many people turn the temperature up and down all the time. Figure 11-10 is the basic accustat. The principle of its workings are extremely simple. A vial of mercury similar to a mercury stick thermometer is used across a set of open contacts. The mercury completes the circuit when it reaches both contacts. These are non-adjustable and are purchased at specific temperature ratings. If the owner wants 76 degrees F. in his establishment, that's it. If he wants to change that, he will need a new accustat sensor. These accustats come in different models from the basic single-stage cooling to a two-stage cooling,

115

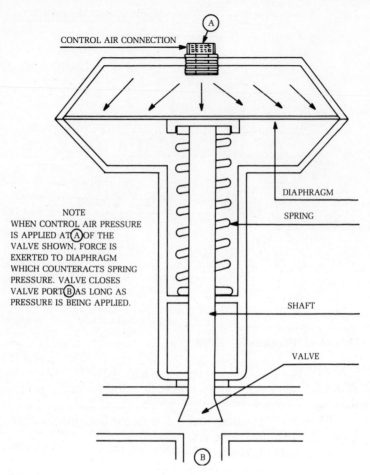

CONTROL AIR CONNECTION

A

DIAPHRAGM

SPRING

NOTE
WHEN CONTROL AIR PRESSURE
IS APPLIED AT (A) OF THE
VALVE SHOWN, FORCE IS
EXERTED TO DIAPHRAGM
WHICH COUNTERACTS SPRING
PRESSURE. VALVE CLOSES
VALVE PORT (B) AS LONG AS
PRESSURE IS BEING APPLIED.

SHAFT

VALVE

B

Fig. 11-9. Pneumatic operated valve.

two-stage heating with an automatic changeover that needs no attention to maintain a set temperature regardless of the season. An embossed numeral on the glass envelope of the sensor designates the temperature of the switch closing.

AMBIENT THERMOSTATS

These are used to keep units from cycling on in cold weather or to place hot water or steam into a water tower during cold weather operations. It is a sensing device for piloting another control that is set to ambient temperature.

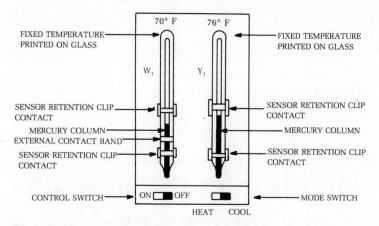

Fig. 11-10. Accustat, with cover removed.

CHAPTER 12

Optional Components in Condensing Unit

In many locations where air conditioning systems have been installed in sub-division homes, or condominiums, or even in strip-stores, the basic straight cool system has very few controls or devices to protect the system. It is for that reason, the price of the unit is kept attractive to the building contractor. The only safety control on many of those units is sheer luck. In this chapter, I want to mention some of the devices and components that are used in more expensive units to create better and longer operation without major breakdowns.

CRANKCASE HEATER

The crankcase heater is a resistive heater that either mounts around the bottom of the body, or is an insertion type that slips into a well in the body, or a bar type that fastens to the bottom of the crankcase. It is the function of this component to energize on the compressor off cycle and transmit heat into the crankcase. The reason for this is to boil any liquid refrigerant that may migrate into the compressor. If an evaporator coil is higher than the compressor at the end of a cycle and the compressor stops, liquid might drop causing it to collect on pistons or in the crankcase of the compressor. On some cool days, refrigerant can condense in the compressor if it isn't being used. The crankcase heater prevents a compressor from trying to compress liquid refrigerant, thus resulting in mechanical failure. The heater is usually wired through a relay that opens the circuit when the compressor is in operation. Some manufacturers leave the crankcase heater energized all the time by wiring it to the line side of the compressor contactor. This can cause a compressor to become too hot when operating, thus opening the overload relay and stopping the compressor operation.

RECEIVER

This refrigerant reservoir called the receiver is found in units that are using a TEX (temperature expansion) valve. When the heat load demand diminishes, the valve throttles the refrigerant flow. The excess refrigerant is stored in the receiver until a load demand is made. In some newer units engineers have designed some of the condenser coils slightly larger than needed to act as a storage area for excess refrigerant. In most light commercial refrigeration systems, along with heat pumps, the receiver is commonly seen. Figure 12-1 shows a typical receiver. This component is installed in the liquid line close to the condenser coil.

ACCUMULATOR

The accumulator is a very important component in heat pumps and low temperature refrigeration equipment. It is a vessel that

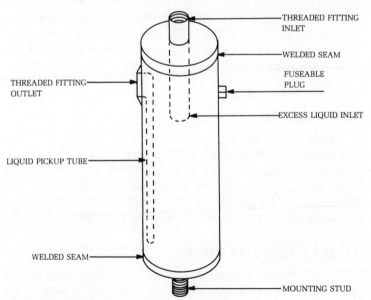

Fig. 12-1. Vertical receiver.

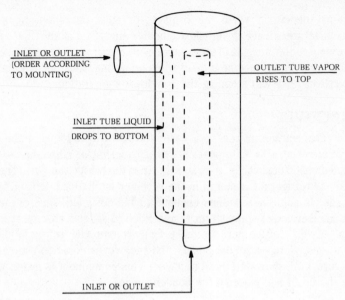

INLET OR OUTLET
(ORDER ACCORDING
TO MOUNTING)

OUTLET TUBE VAPOR
RISES TO TOP

INLET TUBE LIQUID
DROPS TO BOTTOM

INLET OR OUTLET

Fig. 12-2. Accumulator.

collects any liquid refrigerant that is not boiled off in the evaporator. This prevents the compressor from slugging, (trying to compress a liquid) and causing major damage. In some extreme cases, a pipe-heater tape is wrapped around the accumulator and insulated. This ensures the refrigerant will boil sufficiently and enter the compressor as a vapor. Figure 12-2 is a typical accumulator showing how it works. This component is mounted in the suction line just before the compressor.

HOT GAS MUFFLERS

Pulsations created by the pistons of a reciprocating compressor can create objectionable noises. When noise level is an important consideration, or hot gas lines are long, mufflers can be used to minimize the transmission of "hot gas pulsations." Figure 12-3 is an illustration of a hot gas muffler. This component is installed on the discharge line as close to the compressor as possible.

DRIERS-DEHYDRATORS

Driers should be used on both the liquid line and the suction line. The driers are designed differently and should be used for the purpose designated. The liquid line drier filters as well as dries the

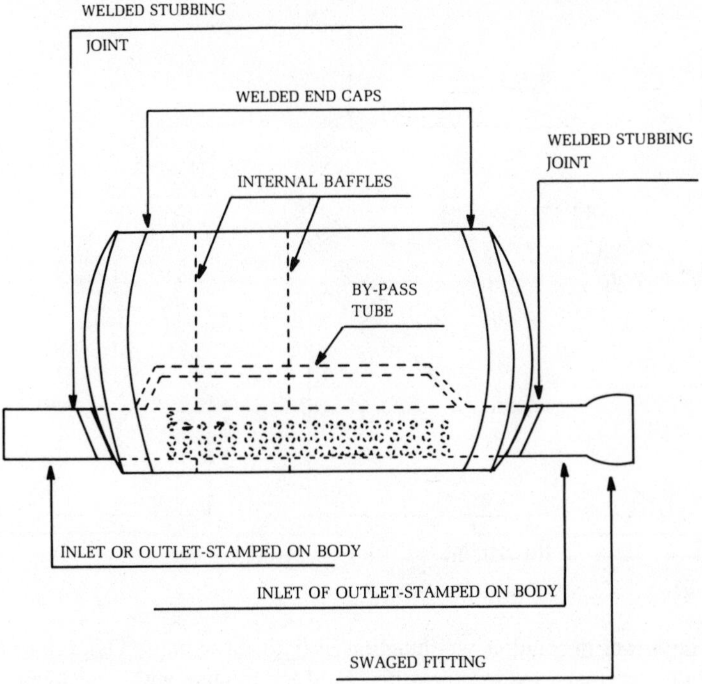

Fig. 12-3. Hot gas muffler.

refrigerant as it leaves the condensing unit. The drier contains a desiccant and a strainer or sieve. The refrigerant flowing through the drier gives up any moisture it may be carrying to the desiccant. Any particles of a solid will be trapped and held in the drier. Most of these units are designed with a two-pound pressure drop across it. When a drier starts to become clogged, the pressure drop increases as a first indicator. Frost might form on the drier when it becomes clogged. The primary purpose of the liquid line driers is the protection of the metering device against dirt and moisture. No more than one liquid line drier should be installed in the liquid line at one time. If two or more are placed in series, the pressure drop could be severe enough to impede refrigerant flow. It is recommended that driers be installed on a vertical plane. The reason is that if the refrigerant charge becomes low, it would only flow over a small amount of the desiccant if the drier were mounted in the horizontal plane. In Fig. 12-4 a drier is shown. Notice the directional arrow that designates the direction it should be installed in the liquid line. If the drier is a bi-flow type, the directional mounting is not

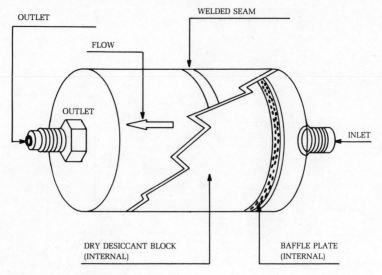

OUTLET

FLOW

WELDED SEAM

OUTLET

INLET

DRY DESICCANT BLOCK
(INTERNAL)

BAFFLE PLATE
(INTERNAL)

Fig. 12-4. Filter drier.

important because it will function in both directions. This type of bi-directional flow drier was designed for the use with heat pumps due to the reverse flow of the liquid when in cooling or heating mode.

The suction line driers are bigger and usually have a Schraeder-type valve on the inlet side of the drier. By taking the pressure at the suction of the compressor and at the inlet of the drier, the pressure drop is easily found. The basic construction of this filter drier is the same as the liquid line. These filters are used to protect the compressor from contamination by acid, metal filings, etc.

Sizing

Both types are made for installation with either a flare or solder. They are sized internally by the amount of cubic inches of desiccant volume. The pipe size is given in eighths of an inch. The beginning numbers are the manufacturer part classification number followed by the sizing digits. A filter drier carrying the last numbers of 083-S designates an eight cubic inch drier with a ⅜ fitting for soldering. If the designation was 165 SAE, you have a 16-cubic inch, ⅝ fitting for flare. These units are in-line filters. On larger equipment found in the light commercial installations a shell-and-core-type drier filter could be used.

122

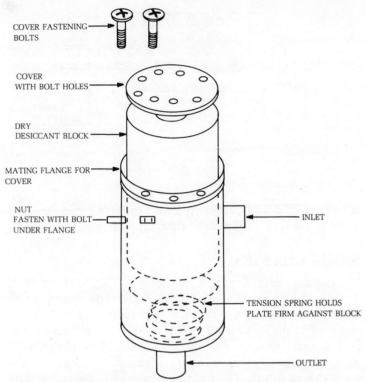

COVER FASTENING BOLTS

COVER WITH BOLT HOLES

DRY DESICCANT BLOCK

MATING FLANGE FOR COVER

NUT FASTEN WITH BOLT UNDER FLANGE

INLET

TENSION SPRING HOLDS PLATE FIRM AGAINST BLOCK

OUTLET

Fig. 12-5. Shell-and-core type of replaceable filter drier.

REPLACEABLE-CORE, SHELL-TYPE FILTER DRIER

In Fig. 12-5 this type of filter drier is shown. The cost of the initial installation of this type of filtration is more expensive than the in-line filter driers. The advantage in this unit is its size, and type of filtration core used. These units are usually stocked in the local parts warehouses in a shell size that will accommodate from one to four cores. The cores are stocked in 48-cubic inch and 100-cubic inch capacity. You can see that you can get as much as 400-cubic inch capacity in one filter drier. I mention it only for you to be aware of them.

STRAINERS

This unit shown in Fig. 12-6 is used mostly on systems using capillary tube metering. It contains a very fine mesh sieve that

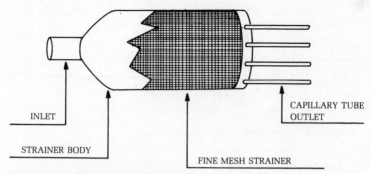

INLET

CAPILLARY TUBE
OUTLET

STRAINER BODY

FINE MESH STRAINER

Fig. 12-6. Strainer filter.

prevents particles of foreign matter in the liquid refrigerant from
entering the very small orifice of the capillary tube.

SIGHT GLASSES

There are several types of sight glasses. They are really liquid
indicators that show the condition of the liquid in the system. Some
of these indicators have a moisture indicator located in the glass.
The color of the indicator changes if there is any moisture in the
system. Different manufacturers use different colors. Green in one
is dry, if it turns to yellow, moisture is present. In another the blue
might change to red. The thing to remember is the use of sight
glasses are usually restricted to equipment using the TEX valve.
Figure 12-7 shows a typical sight glass.

SOLENOID VALVES

The solenoid valve is an electrically-operated valve. It uses an
electromagnetic field to raise a pilot valve. The gas pressure raises

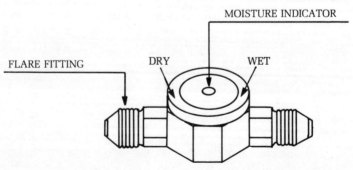

MOISTURE INDICATOR

FLARE FITTING   DRY   WET

Fig. 12-7. Sight glass.

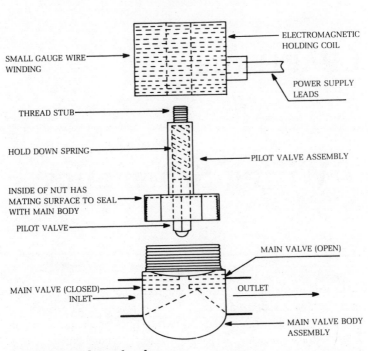

LOCK NUT

ELECTROMAGNETIC
HOLDING COIL

SMALL GAUGE WIRE
WINDING

POWER SUPPLY
LEADS

THREAD STUB

HOLD DOWN SPRING

PILOT VALVE ASSEMBLY

INSIDE OF NUT HAS
MATING SURFACE TO SEAL
WITH MAIN BODY

PILOT VALVE

MAIN VALVE (OPEN)

MAIN VALVE (CLOSED)
INLET

OUTLET

MAIN VALVE BODY
ASSEMBLY

Fig. 12-8. Solenoid valve.

the main valve allowing flow. The flow will continue as long as the
pilot valve is held off its seat. The valve in Fig. 12-8 is typical. It
can be used with liquid as well as vapor or high-pressure gases. In
the condensing unit, it can be used for a pump-down liquid-line valve
or an anti-slug valve. If used for a pump-down valve, the unit must
have a low-pressure control to stop the compressor operation. For
use as an anti-slug valve, the valve is wired to close as soon as the
compressor stops. It prevents any liquid in a vertical liquid line from
dropping down into the compressor. When mounting a solenoid valve,
it is good practice to orient the valve for operation in a vertical plane.
This would require the inlet and outlet of the valve be in a horizontal
plane. The electromagnetic coils are available in 24, 120, and 240
volts.

There are many other component parts that are used in heavy
commercial and industrial equipment condensing units. Those of you
that work with that phase of air conditioning and/or refrigeration know
of the catalogs that can be consulted for a specific application.

CHAPTER 13

Wiring Schematics

I have said many times, "how can you repair something if you don't know what it is supposed to do." For this reason most equipment manufacturers attach a wiring diagram to their products. Some of these diagrams will show the location of electrical components where they are actually placed in the unit. It is then called a component location diagram, and a sample is shown in Fig. 13-1.

Also included in the unit will be a wiring diagram called a "schematic." This diagram shows the actual sequence of operation when the control circuit and high voltage circuit becomes energized. In Fig. 13-2, a sample wiring schematic is shown. Some people become frightened and somewhat lost when they must use a schematic. It is similar to using a road map when making an automobile trip. As on the map, there are symbols that represent real objects. The same is true with the schematic. In fact, just like the map, there is a legend of symbols that might not be familiar to the technician. Many of the symbols used are standardized in the industry. The most common schematic symbols are shown in Fig. 13-3.

The transformer separates the two parts of the circuit, the high voltage being kept on the primary side windings and the low voltage kept on the secondary side windings. The symbol used for the transformer is standard and used by most manufacturers. The voltages will be labeled for identification.

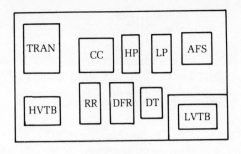

Fig. 13-1. Component location diagram in control panel.

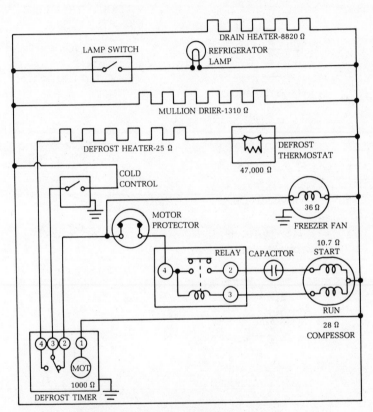

Fig. 13-2. Simple wiring schematic.

I want to explain the low-voltage circuit schematic as shown by Fig. 13-4. Secondary wiring is labeled T_1 and T_2. The T_2 of the transformer is wired to one side of every 24 volt control coil. This is a direct hookup from transformer to control. The T_1 side of the transformer is routed through all of the controls. Safety controls can

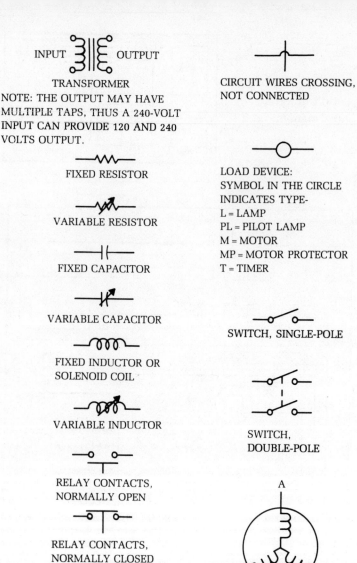

INPUT | OUTPUT

TRANSFORMER

NOTE: THE OUTPUT MAY HAVE
MULTIPLE TAPS, THUS A 240-VOLT
INPUT CAN PROVIDE 120 AND 240
VOLTS OUTPUT.

FIXED RESISTOR

VARIABLE RESISTOR

FIXED CAPACITOR

VARIABLE CAPACITOR

FIXED INDUCTOR OR
SOLENOID COIL

VARIABLE INDUCTOR

RELAY CONTACTS,
NORMALLY OPEN

RELAY CONTACTS,
NORMALLY CLOSED

THERMOSTAT

CIRCUIT CONNECTIONS

CIRCUIT WIRES CROSSING,
NOT CONNECTED

LOAD DEVICE:
SYMBOL IN THE CIRCLE
INDICATES TYPE-
L = LAMP
PL = PILOT LAMP
M = MOTOR
MP = MOTOR PROTECTOR
T = TIMER

SWITCH, SINGLE-POLE

SWITCH,
DOUBLE-POLE

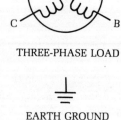

THREE-PHASE LOAD

EARTH GROUND

Fig. 13-3. Electrical schematic symbols.

128

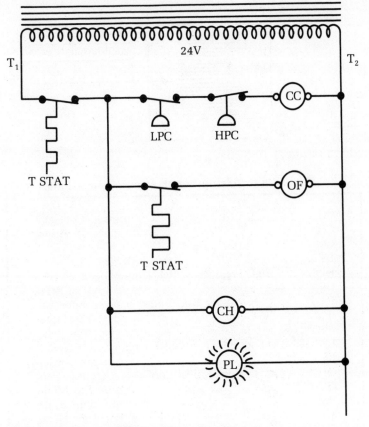

Fig. 13-4. Low-voltage wiring circuit.

also be wired where 24 volts flow through them. The T_1 loop supplies the other side of the coil with the proper voltage for its operation. The holding coil, if energized by low voltage, will appear on this low voltage side of the schematic. The switching action, if controlling line voltage, will be located on the line voltage side of the schematic wiring diagram. The switching action of the thermostat usually occurs in this section of the schematic. Picture in your mind, that everything in this section is low voltage.

On the other side of the transformer, line voltage is being supplied to motors and controls through relays or/and switches. The diagram is similar to the low voltage side with the difference being that high voltage is being controlled. On this side of the schematic, L_1 and L_2 are used in the case of a single-phase unit. If the unit is

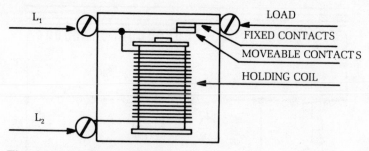

Fig. 13-5. Lock-out relay.

Table 13-1. Wiring Schematic Abbreviations.

CC...........................Compressor Contactor
CH...............................Crankcase Heater
CR................................Control Relay
DT...........................Defrost Thermostat
HPS......High Pressure Switch (control-open on rise)
HR................................Heat Relay
IFC...........Indoor Fan Contactor (evaporator fan)
IFM..........................Indoor Fan Motor
IFR..........................Indoor Fan Relay
LPS.......Low Pressure Switch (control-open on fall)
LS................................Limit Switch
OFC..........Outside Fan Contactor (condenser fan)
OFM..........................Outside Fan Motor
PCB.....................Printed Circuit Board
TB....................Terminal Board (or block)
TDR..........................Time Delay Relay
CB................................Circuit Breaker

three-phase, the L_3 would also be used. Notice that L, which is really the line side of a switch or control will change to T_1, T_2, when it crosses over to the load side of the control or switch. This happens only when the switch is closed.

In most servicing diagnosis, malfunctions will be very easy to detect. Problems like defective motors, blades, capacitors, compressors etc., can be seen. It is the unseen malfunctions that might require you to understand the electrical sequence of the unit. In Fig. 13-5 a simple lockout relay is shown. You can see the path of electricity when the relay is in either position, open or closed contacts.

130

CHAPTER 14

Properties of Refrigerants

In this chapter only refrigerants that will be encountered by service technicians or homeowners will be discussed. Those refrigerants used in heavy commercial and industrial applications will not be covered. I would like to begin by correcting those that have been calling the refrigerant freon, frezone, etc. Every manufacturer has a brand name on his product, and that is the explanation for the name Freon. It was the first brand name to be presented on the market, so that is why most call it that. When I say refrigerant, it will mean Freon, Isotron, Genertron, Carrene, etc.

Comparing refrigerants with the coolants of an internal combustion engine has been done for years because both provide heat transfer. In the case of an automobile, the coolant is circulated through the system with the use of a water circulating pump. The cooled water is pumped into the cooling jacket of the motor block where it absorbs heat from the metal and transports it to the radiator. The radiator has many tubes in it that have fins attached to them. This gives more surface for heat to be transferred to the air passing over the fins. This air supply is usually forced across the fins with the use of a fan. By using a pressurized radiator cap, the boiling temperature is raised. Coolant is lost at times due to leaks or evaporation.

REFRIGERANTS

In the refrigeration system, a refrigerant is used instead of the coolant. The refrigerant is circulated through the system with the use of a circulating pump called a compressor. The compressor pumps the refrigerant into a coil that is located inside the structure. In the coil, the refrigerant absorbs heat and changes state into a vapor. This vapor is brought back to the condenser which is similar to the automotive radiator. In the condenser coil, the heat is transferred to the air in the same way the coolant does in the automobile, and the vapor again changes back to a liquid.

The big difference is that the refrigerant is contained in a sealed system and never has to be replaced or changed unless a leak in the system develops. In order to achieve certain low temperature protection against an engine block freeze up, the quantity of coolant is increased. In the field of refrigeration and air conditioning, different refrigerants are used for different applications.

R-12: CCl_2F_2 (Dichlorodifluoromethane)

This refrigerant was one of the first developed and was used in early air-conditioning systems. It was used in both window units and central air-conditioning systems. It was slowly phased out of the air-conditioning line but continues to hold its position in the refrigeration line. Equipment to handle R-12 was large, and this brought the advent of R-22 into the industry. The most important feature that R-12 has to its credit is its ability to mix with refrigeration oil at low temperatures. Other refrigerants will not mix well with the refrigeration oils at low temperatures. This, was one of the reasons that this particular refrigerant is still in use today. With today's technology, R-502 has been introduced into the low temperature refrigeration market. With this refrigerant equipment size has been reduced.

R-22: $CHClF_2$ (Monochlorodifluoromethane)

This is the most popular refrigerant used in air conditioning systems. From small to very large tonnages, R-22 is still used. Some might be asking why the heck I am giving you the full names of these refrigerants. You'll probably never need them except when playing some type of quiz game. I don't want you to ever say that I didn't tell you. R-22 is a poor refrigerant for low temperature refrigeration. It separates from the oil too easily at low temperatures.

R-500: CCl$_2$F$_2$ 73.8% (Dichlorodifluoromethane)
CH$_3$CHF$_2$ 26.2% (Difluoroethane)

This blend of two components is called an Azeotrope. This is not seen very often in the residential units anymore. There was a time when a certain manufacturer used this refrigerant in their own equipment. They have been phased out except for those that still might be in the field. It was only used in air conditioning equipment.

R11: CCl$_3$F (Trichloromonofluoromethane)

This is used in cleaning the inside of a system. It has a higher boiling point (74.7 degrees F. at 1 atm) than the other refrigerants and acts as a solvent when circulated through a closed refrigeration system. It is used usually when a compressor has had a burn-out. Particles of the motor winding insulations along with other foreign materials can be flushed from the system with the use of R-11.

COLOR CODING

Color coding of refrigeration cylinders (drums) works very well. It enables you to identify a refrigerant at a glance. The homeowner can also be alert as to whether a drum of refrigerant being carried to his air conditioning unit is the right one for his unit. Typical color codes are:

R-12	White
R-22	Green
R-502	Purple
R-500	Yellow
R-11	Orange

TEMPERATURES

Refrigeration is divided into three temperature zones. Equipment and systems are designed with this factor in mind. High-Temperature Zone covers the air-conditioning spectrum and all equipment that operates above 32 degrees F. Medium-Temperature Zone equipment operates between 32 degrees F. and 0 degrees F. Low-Temperature Zone equipment operates from 0 degrees F. to every temperature below zero. These units are designed to function at specific temperatures and trying to interchange them sometimes turns out to be more trouble than its worth. I used to dread when a restaurant owner would ask "How much would you charge to

Table 14-1. Temperature/Pressure Chart.

TEMP F	TEMP C	R11	R12	R22	R500	R502	R717
−30	−34.4	27.8	5.5	4.9	1.2	9.4	
−28	−33	27.7	4.3	5.9	.1	10.5	.0
−26	−32	27.5	3.0	6.9	.9	11.7	.8
−24	−31	27.4	1.6	7.9	1.6	13.0	1.7
−22	−30	27.2	.3	9.0	2.4	14.2	2.6
−20	−29	27.0	.6	10.2	3.2	15.5	3.6
−18	−28	26.8	1.3	11.3	4.1	16.9	4.6
−16	−27	26.6	2.1	12.5	5.0	18.3	5.6
−14	−26	26.4	2.8	13.8	5.9	19.7	6.7
−12	−24	26.2	3.7	15.1	6.8	21.2	7.9
−10	−23	26.0	4.5	16.5	7.8	22.8·	9.0
−8	−22	25.8	5.4	17.9	8.8	24.4	10.3
−6	−21	25.5	6.3	19.3	9.9	26.0	11.6
−4	−20	25.3	7.2	20.8	11.0	27.7	12.9
−2	−19	25.0	8.2	24.4	12.1	29.4	14.3
0	−18	24.7	9.2	24.0	13.3	31.2	15.7
2	−17	24.4	10.2	25.6	14.5	33.1	17.2
4	−16	24.1	11.2	27.3	15.7	35.0	18.8
6	−14	23.8	12.3	29.1	17.0	37.0	20.4
8	−13	23.4	13.5	30.9	18.4	39.0	22.1
10	−12	23.1	14.6	32.8	19.7	41.1	23.8
12	−11	22.7	15.8	34.7	21.2	43.2	25.6
14	−10	22.3	17.1	36.7	22.6	45.5	27.5
16	−9	21.9	18.4	38.7	24.2	47.7	29.4
18	−8	21.5	19.7	40.9	25.7	50.1	31.4
20	−7	21.1	21.0	43.0	27.3	52.5	33.5
22	−6	20.6	22.4	45.3	28.9	54.9	35.7
24	−4	20.1	23.9	47.6	30.6	57.4	37.9
26	−3	19.7	25.4	50.0	32.4	60.0	40.2
28	−2	19.1	26.9	52.4	34.2	62.7	42.6
30	−1	18.6	28.5	54.9	36.0	65.4	45.0
32	0	18.1	30.1	57.5	37.9	68.2	47.6
34	1	17.5	31.7	60.1	39.9	71.1	50.2
36	2	16.9	33.4	62.8	41.9	74.1	52.9
38	3	16.3	35.2	65.6	43.9	77.1	55.7
40	4	15.6	37.0	68.5	46.1	80.2	58.6
42	5.6	15.0	38.8	71.5	48.2	83.4	61.6
44	6.7	14.3	40.7	74.5	50.5	86.6	64.7
46	7.8	13.6	42.7	77.6	52.8	90.0	67.9
48	8.9	12.8	44.7	80.8	55.1	93.4	71.1
50	10	12.0	46.7	84.0	57.6	96.9	74.5
52	11	11.2	48.8	87.4	60.1	101	78.0
54	12	10.4	51.0	90.8	62.6	104	81.5
56	13	9.6	53.2	94.3	65.2	108	85.2
58	14	8.7	55.4	97.9	67.9	112	89.0
60	16	7.8	57.7	102	70.6	116	92.9
62	17	6.8	60.1	105	73.5	120	96.6
64	18	5.9	62.5	109	76.3	124	101
66	19	4.9	65.0	113	79.3	128	105
68	20	3.8	67.6	117	82.3	132	110
70	21	2.8	70.2	121	85.4	137	114
72	22	1.6	72.9	126	88.6	141	119
74	23	.5	75.6	130	91.8	146	123
76	24	.3	78.4	135	95.1	150	128
78	26	.9	81.3	139	98.5	155	133
80	27	1.5	84.2	144	102	160	138

82	28	2.2	87.2	148	106	165	144
84	29	2.8	90.2	153	109	170	149
86	30	3.5	93.3	158	113	175	155
88	31	4.2	96.5	163	117	180	160
90	32	4.9	99.8	168	121	186	166
92	33	5.6	103	174	125	191	172
94	34	6.3	107	179	129	197	178
96	35	7.1	110	185	133	203	184
98	36	7.9	114	190	137	208	191
100	37	8.8	117	196	141	214	197
102	39	9.6	121	202	146	220	204
104	40	10.5	125	208	150	227	211
106	41	11.3	129	214	155	233	217
108	42	12.3	132	220	159	239	225
110	43	13.2	136	226	164	246	232
112	44	14.2	141	233	169	252	240
114	45	15.1	145	239	174	259	248
116	46	16.1	149	246	179	266	255
118	47	17.2	153	253	184	273	264
120	49	18.2	158	260	189	280	272
122	50	19.3	162	267	195	288	280
124	51	20.5	167	274	200	295	289
126	52	21.6	171	282	206	303	
128	53	22.8	176	289	212	310	
130	54	24.0	181	297	217	318	

convert my freezer into a cooler?'' Having been down that road, I always advise technicians to walk away from the request and just tell the customer they are too busy.

Although fluorinated-hydrocarbon refrigerants are comparatively safe to work with and handle, there are a few things you should keep in mind. The refrigerants that you will be handling boil at an average of 30 degrees below zero. If in a liquid state the refrigerant touched any part of your skin, damage would result. Caution must be taken to protect yourself from coming in contact with any liquid refrigerant. Secondly, these refrigerants are heavier than air. This means they will displace air at low levels near the floor and then start to fill the room as if it were a jar. For this reason there must be plenty of ventilation. The refrigerant is colorless, tasteless, and odorless, which means the only indication that you are breathing refrigerant vapors is a state of feeling heavy and getting sleepy as if you just ate a big meal. Always be aware of the height a refrigerant will rise to. Make sure there is ventilation, especially if you are soldering. The third caution is to make sure there is no refrigerant vapor present in the area where you are going to solder with an open flame. The compound breaks down when exposed to an open flame. It becomes phosgene, a very toxic, caustic, pungent gas, which can kill. It will cause a choking and a difficulty in breathing with a burning sensation in the lungs. If you suspect this situation, leave the area

Table 14-2. Conversion Chart: Celsius to Fahrenheit.

Fahrenheit to Celsius			
−30	−34.44	50	10
−28	−33.33	52	11.11
−26	−32.22	54	12.22
−24	−31.11	56	13.33
−22	−30	58	14.44
−20	−28.89	60	15.56
−18	−27.78	62	16.67
−16	−26.67	64	17.78
−14	−25.56	66	18.89
−12	− 24.44	68	20
−10	−23.33	70	21.11
−8	−22.22	72	22.22
−6	−21.11	74	23.33
−4	−20	76	24.44
−2	−18.89	78	25.56
0	−17.78	80	26.67
2	−16.67	82	27.78
4	−15.56	84	28.89
6	−14.44	86	30
8	−13.33	88	31.11
10	−12.22	90	32.22
12	−11.11	92	33.33
14	−10	94	34.44
16	−8.89	96	35.56
18	−7.78	98	36.67
20	−6.67	100	37.78
22	−5.56	102	38.89
24	−4.44	104	40
26	−3.33	106	41.11
28	−2.22	108	42.22
30	−1.11	110	43.33
32	0	112	44.44
34	1.11	114	45.56
36	2.22	116	46.67
38	3.33	118	47.78
40	4.44	120	48.89
42	5.56	122	50
44	6.67	124	51.11
46	7.78	126	52.22
48	8.89	128	53.33
		130	54.44

until it is well ventilated. Common sense in handling of the refrigerant is all that is really needed. Don't take unnecessary chances.

Packaging of refrigerants is also important. As with most commodities, volume buying reduces purchase price. The same holds true for refrigerants. In the automotive industry R-12 is packaged in one-pound cans. In our industry, the refrigerants are packaged in 10, 25, 30, 50, and 157-pound quantities. If you are a technician and working on roof top units alone without a helper, you realize how heavy the refrigerant containers can become. The 30-pounder was very convenient to carry up a ladder or pull to a roof with a rope.

Always confirm what refrigerant is in the system before placing a charge into it. This is another piece of information that can be found on the data plate that was explained earlier. If the data plate is no longer attached to the unit, check the metering device. The device might be stamped or labeled as to what refrigerant is being used. Another way to identify what refrigerant is being used in a system is to take a pressure reading and a temperature reading of the ambient that surrounds the refrigerant. Table 14-1 is a temperature/pressure table. With it you can cross reference to find a refrigerant pressure or temperature. The method is so simple that many service technicians forget about it. For instance, it is a good practice to check the calibration of the gauges on the service charging manifold. All that has to be done is hook your manifold and gauges to a drum of refrigerant. Check the temperature of the air around the drum and see what pressure should be exerted by that refrigerant at that temperature.

If the wrong refrigerant is introduced to the system it will not function properly. Head and back pressure might be out of specified areas. Amperage draw of the compressor might also be out of its proper operating ranges. In some areas, temperature is given in centigrade instead of Fahrenheit. Table 14-2 is a conversion table that can be used in conjunction with the temperature/pressure tables.

CHAPTER 15

Basic Mechanical Refrigeration System

I would like to explain, "the refrigeration effect" before going any further. First, this theory applies to air conditioning and refrigeration. In its most simplest form, refrigeration effect is the moving of heat from an objectionable space to an unobjectionable space. This is true in both heating and cooling. Heat is a form of energy and can only be moved from place to place; it cannot be destroyed.

There are different types of systems that are used to accomplish this refrigeration effect. Absorption, electromagnetic, and mechanical are the most popular. The system used in light commercial and residential mainly is the mechanical type of system. The specific system is known as reciprocal mechanical system. A reciprocal compressor was covered in Chapter 3.

The mechanical reciprocal refrigeration system operates with two pressure zones. The refrigerant operates in a high pressure state and a low pressure state. The extent of the pressure depends upon the specific refrigerant being used in the system. Again we have some rule-of-thumb pressure ranges. In a refrigeration system operating with a 40 degrees F differential, evaporator coil approximate pressures would be,

R-22   69 psi on the low pressure side of the system
          225 psi on the high pressure side of the system
R-12   37 psi on the low pressure side of the system
          125 psi on the high pressure side of the system

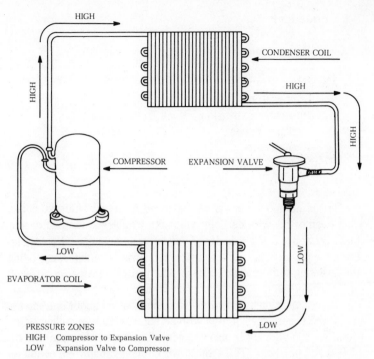

Fig. 15-1. Simplified mechanical reciprocal refrigeration system.

In the automobile air conditioner, R-12 is used as the refrigerant. The operating pressures of R-12 are different in this application. Remember that the compressor is now operating at variable speeds, depending upon the speed of the automobile engine. For this reason, the tonnage of the system changes. The faster the compressor pumps, the more capacity the unit has. Low-side pressure will then vary from approximately 10 to 30 psi and the high pressure side will vary from approximately 125 to 300 psi.

Figure 15-1 shows the simplified mechanical reciprocal refrigeration system. The drawing shows the two pressure zones and where the pressure changes. The four major components shown are:

Compressor
    Hermetic
    Semi Hermetic
    Open Drive

139

Condenser
    Air Cooled
    Water Cooled
    Geothermal Cooled
Metering Devices
    Temperature Expansion Valves
    Automatic Expansion Valves
    Capillary Tubes
    Accumulator
Evaporator Coil

These components will be explained in detail. Understanding what each of these units does will make troubleshooting an easier task. Systems can be designed for air conditioning and/or refrigeration equipment with the same components. The difference being the resulting temperature. This is achieved with proper controlling of the refrigerant.

In air conditioning, systems are usually sized about one ton for every 400 square feet of area with an eight to 10 feet ceiling. One ton of air conditioning is also considered to be one horsepower or 12,000 Btu per hour. The system will circulate about one pound of refrigerant per ton per minute. This circulation is accomplished with the use of the compressor.

COMPRESSOR

The compressor is rated according to how much refrigerant is circulated. The rating will vary according to which refrigerant is used. The capacity changes due to the fact that a specific amount of refrigerant must be circulated to accomplish the refrigeration effect. For example, three pounds of refrigerant being pumped per minute would give you three tons of capacity if R-12 were the refrigerant. If R-22 was pumped in the same compressor, the capacity would drop to 2¼ tons. The reason is that four pounds of R-22 would have to be circulated to accomplish the full three tons. The displacement of each cylinder remains the same.

The compressor pumps the superheated gas (high side, discharge line) into the condenser. On the low side (suction line) vapor from the evaporator is brought back to the compressor for re-cycling. The compressor is sometimes called a pump due to the fact it does circulate refrigerants.

CONDENSER (AIR-COOLED)

The condenser is a length of tubing with fins placed perpendicular to the tubing. Figure 15-2 shows a typical air-cooled condenser coil. This type of condensing coil is used in all size units from an average window unit to a commercial system having a three digit tonnage rating. The amount of refrigerant circulated per minute has a direct bearing on the tonnage rating of the system. The tonnage rating of the unit will determine the physical size of the coil in an air-cooled condensing unit. In Fig. 15-2 the affected heat transfer from the coil to the ambient air is shown. You can easily see that the ambient air temperature will determine the heat rejection efficiency of the condenser coil. Air conditioning systems are designed for the areas in which they are to operate. For instance, in a certain geographical area the unit might be designed to maintain a 72-degree F. temperature in a conditioned space, at an ambient temperature of 95 degrees. As the variable ambient changes, so does the temperature of the conditioned space. Not going into the engineering of it, air conditioning systems have a design factor. The perfect operating temperature of a condenser for example would be 105 degrees F. If the ambient temperature begins to drop, the system begins to lose efficiency. In Fig. 15-3 a condenser cross section is shown. The air flowing over the coil must be controlled if the unit is to operate efficiently during low ambient periods.

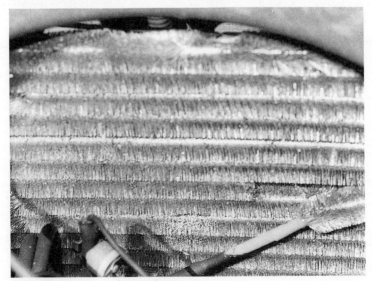

Fig. 15-2. Air-cooled condenser.

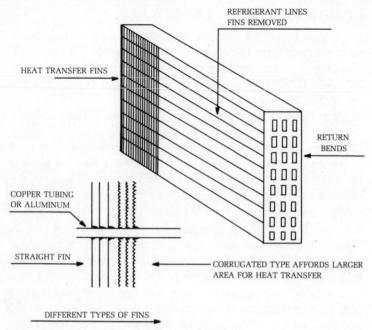

Fig. 15-3. Condenser coil cross-section.

Fan cycling is controlled with the use of a reverse acting high-pressure control. The switch closes on rise of head pressure. The cycling switch might not be enough in those areas where a cold prevailing wind blows through the condenser coil.

In the prevailing wind situation, face dampers are placed on the condenser coil and piped to operate with head pressure. The dampers shown in Fig. 15-4 are used on many commercial applications. The dampers move in proportion to the head pressure. The only time the damper is in its full open position is when the compressor is operating at its design temperature.

This is not important to residential installation, but for the light commercial usage it is. Just think about a store owner with a freezer full of meat, with a walk-in freezer that is inoperable due to a low ambient. Even in light commercial air conditioning this could be important in places like a banquet hall. In those areas where cold weather is a rare occasion, a quick and temporary fix might be a piece of cardboard placed in front of the condenser coil to block the wind; a small hole might then be cut in the cardboard to allow a small amount of air to pass through the coil so it will not get too hot. On the other side of the coin, if things are too hot, the same problem

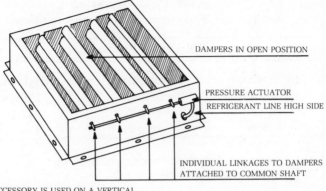

DAMPERS IN OPEN POSITION

PRESSURE ACTUATOR
REFRIGERANT LINE HIGH SIDE

INDIVIDUAL LINKAGES TO DAMPERS
ATTACHED TO COMMON SHAFT

NOTE
THIS ACCESSORY IS USED ON A VERTICAL
OR HORIZONTAL PLANE. INSTEAD OF
THE REFRIGERANT PRESSURE ACTUATOR
BEING USED, A TEMPERATURE CONTROLLED
ELECTRIC MOTOR CAN BE USED.

Fig. 15-4. Face dampers over condenser coil, used in low ambient geographical locations.

can occur. In this case, there could be a defective condenser fan motor that is not available for one reason or another. The proper placement of a water hose and a fine spray directed at the condenser coil can save the day for someone. It is some of these simple things that make the good technician better. I've always instructed students that most of the service call should be done mentally, before picking up a tool. Talking to the owner and thinking carefully during the diagnosis will save time and a lot of your energy. Thinking over the job first will enable the right tools to be selected and brought to the job site. I have watched many technicians almost walk themselves to death, back and forth from the service truck to get the tools they forgot. Pre-planning or pre-thinking can help eliminate the condition.

Refrigeration and air conditioning equipment must operate at their design temperature and pressure, in order to perform their specific job. Table 15-1 shows an example of some operating pressure found in general air conditioning, (high-temperature refrigeration). The equipment listed in the table are air-cooled condensers with a DX (direct expansion) evaporator. I want to say once again that when replacing a part, when it has a specific type of function, the part replacement should be exactly like the part being replaced. This is a very critical point when replacing a fan motor or fan blade. For instance, if a motor that drives the condenser fan blade is being replaced, and the speed is less than the original, the unit might function alright on cooler days. When the ambient temperature rises, the fan

Table 15-1. Operating
Refrigerant Pressures (for design conditions).

OPERATING PRESSURES*		
Refrigerant	**Low Side**	**High Side**
R-22	69 psi	225 psi
R-12	37 psi	125 psi
R-500	46 psi	150 psi
Automotive air conditioning		
Factory installed units		
R-12	10 to 30 psi	125 to 300 psi
Field installed units		
R-12	10 to 37 psi	125 to 300 psi
*Pressure taken while operating at design temperature.		

will not turn fast enough to provide proper heat transfer. This will
cause the unit to be less efficient on the hotter days of the year.
Not only will it drive the owner crazy, it will drive the next service
technician crazy trying to find the problem on the cooler day. There
are certain relays and electrical parts that might be used in a
replacement situation and still perform the same task as the original.
Remember that in certain items within the unit, replacement parts
must be the same. Another example would be replacement of a
section of capillary tubing. If the exact I.D. (inside diameter) is not
available, a different size and the exact length of the original won't
work. If the size is changed, so is the operating pressures which
in turn affect the design factors of the unit.

CONDENSER (WATER-COOLED)

When discussing the air-cooled condensers, it was mentioned
that the ambient air temperature would affect the efficiency of the
unit. The higher the ambient reaches over the design temperature,
the more inefficient the unit becomes. For this reason, water-cooled
units have made their entrance. Of course the water-cooled
condenser has many advantages along with its disadvantages. Some
areas use water-cooled units in residences and light commercial units.
I cover it here for your information.

Water is used as the heat transfer medium in place of air. Water flows through a heat exchanger where refrigerants contained in separate piping give up their heat to the water. Figure 15-5 shows how a few typical condensers are constructed. Now that you've seen the actual condenser and how it is made, you have to get water to make it operate.

In making a selection of whether to use air-cooled or water-cooled, the source of water should be studied very carefully. Water can be secured from a well, city, seawater, brackish, lake, pond, man-made pond. You can see there are many options; however, the content of the water is very important. The reason is that you don't want a constant battle to keep the condenser clean from minerals or rust. A very rapid buildup on the inner surfaces that causes an inefficient heat exchange is costly to maintain. An air-cooled condenser might be advisable in such a case, or perhaps a water treatment system that would be cost effective. In many areas where an apartment complex or a condominium complex uses a community tower, the maintenance fee is not too high if chemical treatment

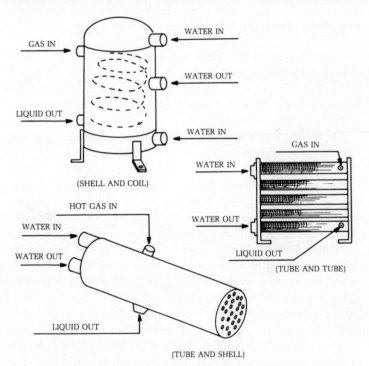

Fig. 15-5. Three different water-cooled condensers.

145

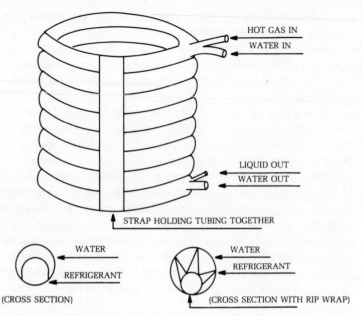

HOT GAS IN
WATER IN

LIQUID OUT
WATER OUT

STRAP HOLDING TUBING TOGETHER

WATER
REFRIGERANT

(CROSS SECTION)

WATER
REFRIGERANT

(CROSS SECTION WITH RIP WRAP)

Fig. 15-6. Tube-and-tube water cooled condenser.

administered through the water tower is done on a regular schedule to keep the individual condensers located in the individual apartments clean. The individual air conditioning units might range in tonnage from two to five tons. It might be necessary for you to be aware of how the water works.

In Fig. 15-6 a tube-in-tube type condenser is shown. It is very common in residential package units. The package can be tucked away neatly in a closet where it hooks up to a common water riser that is usually built into the wall. All the apartments located above and below use the same riser. Each apartment has a set of valves on the main riser in the event the unit must be removed; the water valves can be closed. What happens when a water valve doesn't close completely? Never, never, never cut a pipe until you are sure the valve is holding. If it doesn't, you might be emptying a community water tower all over some expensive carpeting in someone's apartment, as well as that in apartments below. If a valve doesn't hold, it will be necessary to find the shut-off valves for the entire riser. The drain valves somewhere below till then have to be found in order to drain the riser so that new valves can be installed. It is always good policy to try and get all the owners on that riser to agree to renew valves at the same time. You might want to turn the whole thing over to another service company.

In water-cooled condensers, some are sealed and some are built with bolted end plates that enable a technician to service it. The sealed condenser such as the tube-in-tube is cleaned with the aid of an acid circulating pump. Both inlet and outlet sides of the water circulating circuit of the condenser are opened. Fittings are placed on both sides to accommodate water hoses. One end of one hose is placed in the inlet fitting, and the other end of the same hose attaches to the discharge side of the acid circulating pump. Figure 15-7 shows a typical acid circulating pump. The pump is set into a five gallon plastic pail. The other hose is attached to the outlet side of the condenser and the other end of the hose is laid into the bottom of the pail. A gallon of sulfuric acid is placed into the pail and circulated through the condenser. The same acid can be used again, but the acidity should be tested often with litmus paper. If the acid level drops, more acid should be added. In using this cleaning procedure,

MOLDED PLASTIC ENCLOSES PUMP DRIVE MOTOR ENABLING
PUMP TO BE SUBMERGED IN ACID

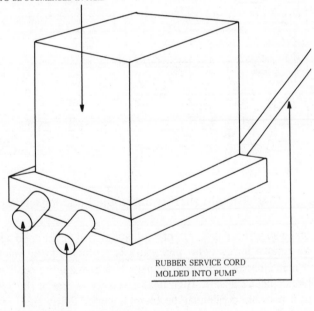

RUBBER SERVICE CORD
MOLDED INTO PUMP

PUMP INLET AND OUTLET HOSE CONNECTIONS

Fig. 15-7. Acid circulating pump used to clean and de-scale a water cooled condenser.

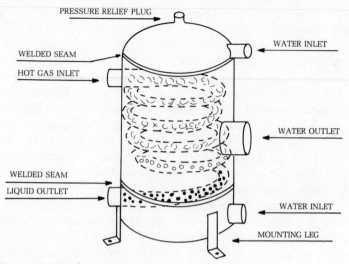

Fig. 15-8. Coil-in-shell water cooled condenser.

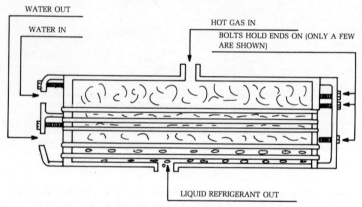

Fig. 15-9. Tube-in-shell water cooled condenser.

make sure there is plenty of ventilation. You might need a small fan to remove fumes. This procedure is continued for about an hour. It helps eliminate mineral deposits that insulate the condenser and hinder heat transfer. High head pressure is one of the symptoms of a possible condenser problem. Figure 15-8 is another type of sealed condenser called coil-in-shell. It too is cleaned the same way the tube-in-tube is cleaned.

Figure 15-9 is a tube-in-shell condenser. This type unit is used in refrigeration and air conditioning equipment from small tonnages up to very large units. Cleaning this type is more difficult to do. This condenser is constructed with end plates that are designed to be removed for cleaning. A long rod is used, something similar to a ram-rod used in rifle cleaning. A wire brush is attached to the end of the rod. With the aid of an electric drill (usually a ¾-hp slow speed drill is used) each tube of the water circuit is cleaned. After the brush is placed through the tube, it is flushed with water. By holding a light at one end and looking down the other (again as in cleaning a gun barrel), one can see mineral or other deposits. The same tube might have to be reamed and flushed with water several times before the deposits break loose.

A quick way to check the efficiency of the condenser is to touch the water outlet side and the refrigerant (liquid) outlet side of the condenser. The refrigerant should feel warm, not hot. The water outlet should also be warm to the touch. If the outlet water pipe is cold, either too much water is flowing through it or there is no heat transfer taking place. If the outlet pipe is very hot, there is a restriction in the water supply.

WATER SOURCE

Due to the fact that you might be called upon to service a residential unit in a complex using one of these water sources it will be covered in this section. In fact you might even get a call on a private home that is using a water-cooled condensing unit.

Once-Through

This type of water supply uses a private well, sea water, lake water, etc. The water is brought into the water-cooled condenser with the use of a circulating water pump. The pump places the water into the condenser and after it circulates one time through all the tubes, the water is returned to the ground with the use of another well usually called a dry well. If the source is outside surface water it is returned to the surface water source. The slight elevation in temperature is not harmful to the environment, in certain areas. Another source is city water. This is only cost effective in an area where the price of city water is very low. In using any raw water supply other than city water, careful attention should be taken as to pre-filtering as not to cause any problems in the condenser coil.

Water Tower

The initial cost of this system might be higher but the long term benefits are high. In the once-through system, water temperature is a constant factor regardless of the ambient temperature. This lower operating pressure helps the unit be more efficient. The cooling tower operates with the same premise. The additional cost is due to equipment. All need the water circulating pump; however, there is a choice of water towers. Again, let me make mention that equipment differs depending upon geographical location.

The first type of tower I want to discuss is probably the most inexpensive, the atmospheric tower pictured in Fig. 15-10. This tower can only be used where there are prevailing winds almost constantly. The construction of the tower is simple. Usually made of redwood, there are some that are being constructed out of plastics. The principle of the tower is to cool the water in the sump by the means of evaporation. The tower has a main water header that brings the water to individual spray nozzles. At the nozzle, the water is broken into fine particles giving the air more surface to absorb heat. The small amount of water that boils off into vapor takes heat from the main body of water, cooling it. The wind must pass through the tower to carry heat-laden moisture away from the area. The slats in the tower are placed in the interlocking sloped position to keep

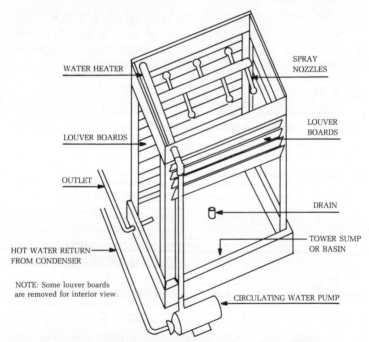

Fig. 15-10. Atmospheric water tower.

the main body of water from being blown out of the tower. A water valve keeps the sump or basin filled to a specific level to make up for that water that is lost through evaporation. The amount is small. The water is pumped from the discharge of the circulating pump to the condenser. From the condenser, the water is pumped to the riser of the tower into the header where it is cycled again to become cool before it falls to the sump below.

Induced-Draft/Forced-Draft

This type of tower operates on the same principle as the atmospheric tower, evaporation. This tower is designed to be used in the event there are not enough prevailing winds. With the aid of a large fan, great amounts of air are drawn into the tower and across the water to cause evaporation. Figure 15-11 shows an induced draft tower.

Certain considerations must be made before selecting a water tower system. You must be in an area where 100% R.H. (relative humidity) is not a frequent occurrence. When air is at 100% R.H., it is saturated and can't absorb any moisture, thus evaporation can't take place rendering a tower useless. There are certain geographical locations where this condition does exist most of the time. The sec-

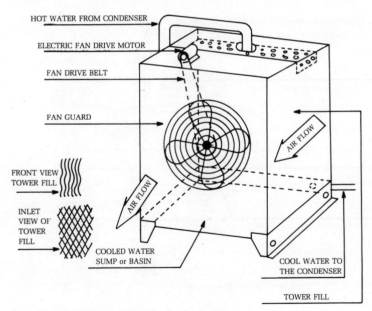

Fig. 15-11. Induced-draft water tower.

ond thing to consider is the maintenance of the tower and water-cooled condenser. Chemical treatment is a must to prevent the buildup of mineral materials and algae. If these are allowed to progress without the use of chemical treatment, they are capable of the total destruction of the equipment, forcing frequent replacement. In some areas where city water is used, the use of a water bleed-off in the return riser is a great help.

Head Pressure

Head pressure in water-cooled units is achieved with the use of several controls. In Fig. 15-12 a pressure-activated water valve is used to maintain constant head pressure. It is an adjustable valve that can be varied for the different refrigerant operating pressures. The spring tension operates against head pressure. This type of valve is very popular in both water tower applications and once-through systems.

Maintaining a uniform water temperature in the tower sump depends on the ambient. An aquastat is used to cycle the tower fan. When the water temperature gets low, the fan stops pulling air through the tower. This in turn lets the water in the sump heat from the return water of the condenser. The second stage of an aquastat

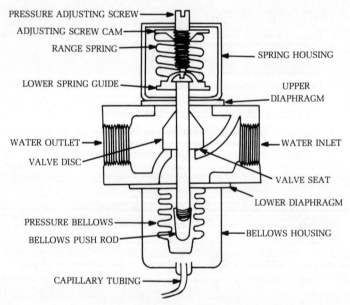

Fig. 15-12. Pressure-controlled water valve.

can energize a water valve that will not allow the water to go to the riser and flow through the tower fill. The valve lets the water return to the sump directly. This is helpful when there are gusty prevailing winds, that cool the water instead of the fan. The aquastat is a watertight unit that usually mounts on the tower itself. The watertight sensing bulb senses the water in the tower sump. In very extreme cases when air conditioning systems must be used in very low ambients, steam or hot water from a boiler are pumped into the tower sump. In some residential applications, the tower might have to be flooded by the make-up water to raise the sump temperature. This is done with the use of a solenoid valve placed before the float valve.

A word of caution is due here. Most of us know that electricity and water don't mix. Bear this in mind when using electric drills for cleaning and repairing. Make sure you have a good grounding device for the tool. Beware of rusted out metal tubing the encases wires that might be laying in water. It is usually the experienced technician that has the accident, not the novice. Technicians sometimes get rushed and forget safety rules.

EVAPORATOR-DX

The evaporator coil is the component in the system that absorbs heat from the conditioned space. It is similar in design to the air-cooled condensing coil. It is also finned to create more area to contact air flowing across it. The evaporator coil is designed for a specific purpose, to control humidity and temperature. Some coils are designed for high de-humidification, others are not. Earlier in the book, I mentioned that perfect conditions would be 72 degrees F., and a 50% R.H. for most human beings. A special design is required to accomplish this. If an evaporator is being used to keep vegetables and fruit from spoiling, a different type of coil is needed for the removal of heat and not the moisture contained in the product. This is only one example of many applications. Each evaporator is different and is designed to perform a specific task.

The refrigerant is pumped into the evaporator coil and pressure is dropped upon entering. The entrance is usually at the bottom of the coil. The refrigerant enters the coil as a liquid and begins absorbing heat. By the time the refrigerant reaches the top rows of the evaporator coil, it should have boiled and turned into a cool vapor. It is this cooled vapor that returns to the compressor to cool its windings and then it is recycled into a liquid to be circulated again.

Boiling temperatures of refrigerants at one (1) atm of pressure.

R-12   −21.62 degrees F.
R-22   −41.4   degrees F.
R-500   −28.0   degrees F.

METERING DEVICES

This is the fourth major component in the refrigeration system. It is at the metering device where the refrigerant held at high pressure in a liquid state passes through, drops in pressure, and sub-cools itself just prior to entering the evaporator coil.

Metering devices fall into four (4) major categories.

Thermostatic Expansion Valves
Automatic Expansion Valves
Capillary Tubes
Accurater (Carrier Device)

Thermostatic Expansion Valve

In Fig. 15-13 a typical, simplified, thermostatic expansion valve is shown. In the type of equipment being serviced, this type of metering device is most effective. By using a device such as this in a piece of equipment, the cost is elevated. The use of this valve

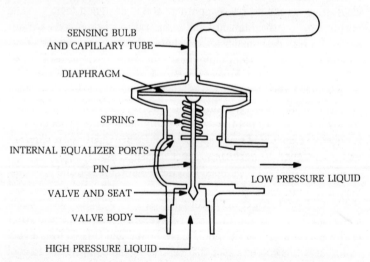

Fig. 15-13. Thermostatic expansion valve.

requires a receiver to be used also. If it cost more to manufacture a unit such as this, why not use a different metering device? The advantage of this type of metering device is its ability to respond to variable load factors. Look at the diagram in Fig. 15-13 again. The sensing bulb is mounted to the suction line at the outlet of the evaporator coil. If the temperature is too high, the valve will open to allow more refrigerant to enter the coil. If the temperature being sensed should become low, the valve will throttle down the amount of refrigerant flowing into the coil. That is the greatest advantage to this type of metering device, it senses gain in heat load and can compensate for it. For example, at a wedding when people are sitting inactively, the heat buildup will be low. If the reception were in the same room, and people started dancing, the heat load would increase. The device used in a commercial type of refrigeration system where people are constantly opening and closing the appliance works very well keeping up with the demand.

Automatic Expansion Valve

This valve has an adjustable orifice to control pressure which in turn controls the temperature. This device is limited to the pressure for which it is adjusted. The only advantage that this type of device has over the capillary tube is that it is adjustable where the capillary tube isn't.

Capillary Tube

This metering device is a very popular type. It is used in both refrigeration equipment, commercial and residential, and air-conditioning equipment, both residential and commercial. The small, copper tubing is sized by its I.D. (inside diameter). The liquid line from the condensing unit might be ⅜-inch copper tubing. When it enters the evaporator section, the liquid line restricts down to a capillary tube with an I.D. of perhaps .065 of an inch. By restricting the refrigerant, pressure is raised on the high side of the system. The length of the tube is also relevant to the pressures. You can see that this system has a fixed metering device. If a large heat load is placed upon this type of unit, it takes a long time to remove the heat load. You can see that the thermostatic expansion valve is far superior in responding to changes in heat load. The capillary tube system works well in home units, both air conditioning and refrigeration. Figure 15-14 shows the typical capillary tube system. Care must be taken to protect the inside of the tube from becoming

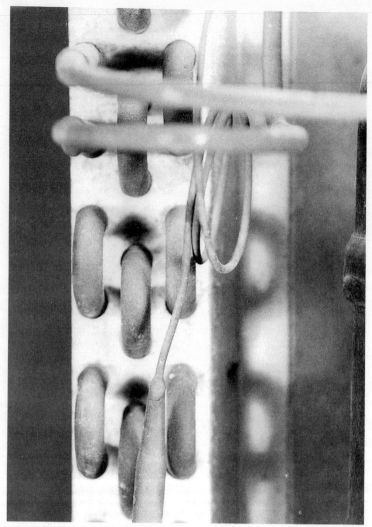

Fig. 15-14. Capillary tube.

restricted. This is the reason for the fine mesh strainer at the entrance of the tube. All foreign particles should be trapped here and contained.

Accurater

This device eliminates the use of the capillary tube. A specially designed body houses a small piston type of device. In one direction of flow, the piston seats against the body and forces the refrigerant to flow through a specially designed orifice. In the reverse flow of the refrigerant, the piston is pushed off its seat in the valve body allowing the refrigerant to flow around the piston without any restriction that might cause an interference with pressure. With an accurater at each end of the unit, a heat pump operates very efficiently.

CHAPTER 16

Humidity Control

In air conditioning and refrigeration a word keeps popping up that some might not be acquainted with; for that reason, I am going to spend a little time explaining **psychrometrics**. The word is defined as the science involving thermodynamic properties of moist air and the effect of atmospheric moisture on materials and human comfort. Don't let the word, or the definition frighten you. It is the job of the engineer to use his slide rule to come up with the answers. The service technician must be aware of what it is, so when he encounters a humidity problem, he will know how to correct it. To begin with, let me give you some of the terms that are used.

Relative humidity—ratio of the actual water vapor pressure of the air to the saturated water vapor pressure of the air at the same temperature.
Dry-bulb temperature—the temperature of air as registered by an ordinary thermometer.
Wet-bulb temperature—the temperature registered by a thermometer whose bulb is covered by a wetted wick and exposed to a current of rapidly moving air.
Dewpoint temperature—the temperature at which condensation of moisture begins when the air is cooled.

The four terms listed above are all you need for now. The humidity factor is very important. It is overlooked by many service technicians.

In a residence, high humidity can cause irritability in the occupants and go so far as cause the wallpaper to fall from the wall. In commercial applications such as manufacturing, humidity control is very important. The food industry depends very heavily upon the proper temperature and humidity for storage of foodstuff.

Human beings are most comfortable when the humidity is 50% R.H. Remember you must consider the area in which you are working. In some areas of the planet, humidity is very low, and in other places the humidity is very high. Knowing the R.H. factor that makes the human body comfortable is the degree a service technician should strive to accomplish. Some of you may choose to become technicians in heavy commercial or industrial air conditioning or refrigeration and having the knowledge of psychrometrics will definitely be a help to you.

The basic instrument for testing humidity is called a sling psychrometer. Figure 16-1 shows the instrument. It is basically two identical thermometers. One is used to record the dry bulb temperature. The other one has a small fabric sock that covers the bulb of the thermometer. The sock is moistened with water, then it is spun in a circular motion by its handle for about a half a minute. Evaporation takes place at the bulb of the thermometer causing a cooling effect. Most of these instruments have a scale printed on them where you align the db (dry bulb) and the wb (wet bulb) to get a R.H. read-out. This is the first step to take in humidity control, finding the humidity condition that exists in the conditioned space.

Working to strive for proper humidity, remember that the cooler air becomes, the less moisture it can hold. The reverse of what would be expressed as, the hotter air gets the more moisture it can hold. Example: if a conditioned space in a factory has reached a temperature of 72 degrees F., but the R.H. is 80%, continual cooling would cause a discomfort to the people working in the conditioned space. For this reason heat is applied to the supply air after it has passed over the evaporator coil. By raising the supply air temperature, the ability for it to absorb more moisture in the conditioned space is increased. This procedure is called, re-heat. The removal of moisture from the air is called dehumidification. The air passing over the cold evaporator coil causes water vapor in the air to condense into liquid form. The condensation is collected in a condensate pan and piped out the condensate drain. Drains must be kept clear, in order to remove the condensate water out of the structure.

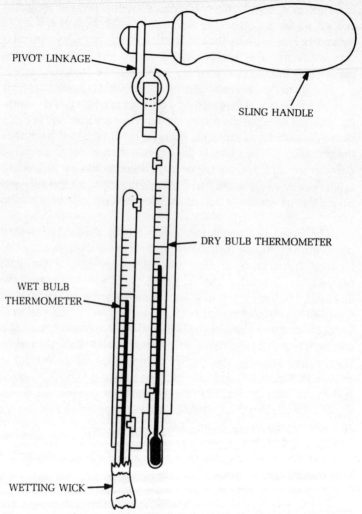

PIVOT LINKAGE

SLING HANDLE

DRY BULB THERMOMETER

WET BULB
THERMOMETER

WETTING WICK

Fig. 16-1. Sling psychrometer.

In other regions, the opposite applies. In an area where the humidity is extremely low, moisture has to be placed into the conditioned space. It can be accomplished by injecting water in mist form into the supply air stream. This is sometimes done with the use of a boiler in industrial applications. If the humidity is 20% R.H., then it has to be raised to 50% R.H. if you are concerned with human comfort. In some instances where an air conditioning system is not being used, a piece of equipment called a humidifier can be used.

The control that is used for this type of work is called a humidistat. This control uses the theory of human hair expanding and contracting in proportion to the amount of moisture contained in the air. The control uses synthetic materials now along with human hair. The expansion or contraction feature of the control operates a switch. This switch is wired to control re-heat when being used for dehumidification and for a water supply valve for humidification.

Figure 16-2 shows a humidistat. When wired for de-humidification, the compressor operates whenever the humidistat switch closes. The circuit is parallel. When the humidity is within the range desired, the compressor will cycle on and off cooling the air. When the humidistat calls for dehumidification, the heat comes on with the compressor to extract the maximum amount of mois-ture from the air. In many residential structures, the energy cost to maintain the ultimate humidity would be very high. For this reason, there is no urgency. If you are working on commercial or industrial equipment, you might have to make adjustments to maintain the proper humidity for the situation.

In a typical supermarket in certain areas the labels on the canned goods will come loose if humidity is not kept within specified limits. Certain computer rooms have problems with humidity causing stoppages. In certain manufacturing, meticulous humidity control has

NOTE
POWER ELEMENT IN HUMIDISTAT IS SENSITIVE
TO HUMIDITY. IT CHANGES LENGTH IN PROPORTION
TO CHANGE IN HUMIDITY.

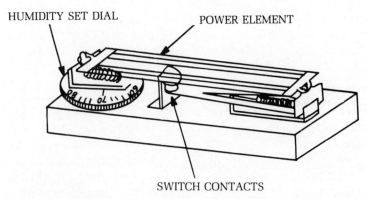

Fig. 16-2. Humidistat, with cover removed.

161

to be observed. Products like cosmetics, explosives, and electronics are also very sensitive to humidity. In food storage, humidity is very important along with the temperature. Produce will be destroyed very quickly if moisture is removed from it.

You can now see the importance in knowing how to control humidity in a controlled environment. In very critical situations a test instrument called a recording psychrometer is used. This instrument will constantly record humidity readings on a circular chart that records for 24 hours or more. Residential customers should be made aware of the expense of trying to control humidity to close ranges.

CHAPTER 17

Basic
Air Conditioning
Operation

I'm going to open this chapter with a few rules of thumb. Usually, 400 square feet of area with an eight-foot ceiling can be cooled with one ton of air conditioning. One tone of air conditioning is equivalent to 12,000 Btu (British thermal unit). One Btu will raise the temperature of one pound of water one degree Fahrenheit.

When a unit is being sized for a structure, many factors have to be considered. Things such as wall thickness, what material is used in their construction. Color of building, amount of glass area, exposure, along with many other factors are considered in the engineering computations for the heat load. In a residence that is simple without any unusual configurations 400 square feet per 12,000 Btu formula can be used. The reason for it being mentioned here is in the event a homeowner has a complaint that his/her unit is not cooling enough, you will notice immediately if a 24,000 Btu unit is trying to cool a 1600-square foot house.

I'm going to discuss a split, straight cool, three-ton system. Placement of the condensing unit is very important. The first thing you should consider is noise factor to the home and neighbors. Some condensing units are more noisy than others. If a unit has objectionable noise, it should not be placed near sleeping quarters. If a replacement unit is being considered, it wouldn't be a bad idea to find one of the intended units operating in the field and listening to any noise that it might generate. Condensing unit fans are designed to discharge air either in a vertical or horizontal plane. With this in mind, when placing a vertical discharge unit you wouldn't place it under a low roof overhang or under a window awning. In the case of the horizontal discharge condensing units, there has to be ample room for air to enter the unit and room for it to exit. If being placed on a roof, consideration to the prevailing winds is prudent.

Depending upon the building codes in your area, the condenser unit should be raised above the ground. There should be freedom of air circulation under the unit to inhibit the formation of rust due to the unit constantly being wet. A couple of $4'' \times 4''$ timbers would lift the condenser off the ground high enough for ventilation to take place.

The air handler (evaporator section) can be located in many locations depending upon the structure. In residences the air handler might be installed in a garage, interior closet, or in the attic. In a straight cool unit or heat pump unit, the air handler contains fan, coil, and controls. If the air conditioning system was added after the heating system was installed, a metal plenum is constructed usually above the heat exchanger of the furnace. The coil and metering device is located in the plenum. The fan of the furnace is used for the air-conditioning unit. The controls are mounted in the furnace section. This can be either a gas or an oil furnace.

There must be an airway for air to get back to the unit. If the evaporator section is located in a closet, a louvered door is sufficient to allow air passage back to the unit. If the evaporator section is located in the garage or in the attic, a return air grill is installed. The grill is an entry to the duct that directs the conditioned air back to the evaporator coil. There should always be a filter to clean the air prior to its entering the evaporator coil. It could be located behind the return air grill on a filter rail, in the ductwork on a rail, or in the evaporator section itself on a rail.

In most of the systems today, the transformer that supplies the low voltage for the control circuit is located in the evaporator section. Few units still place the transformer in the condensing unit. For the wiring of the thermostat, see Chapter 11.

If the evaporator section is located in the attic, or a crawl space above a finished ceiling, an auxiliary condensate drain pan should be installed. In certain regions this is required by building code. The purpose of the auxiliary pan is to collect any water that might spill from the unit in the event the primary drain becomes restricted. If an auxiliary is not used, water can quickly damage the ceiling below the unit.

There are two ways that an auxiliary pan can be used. In most cases, a drain line is attached to the pan and piped to the outside of the structure where it drips in a conspicuous place. Next to the entrance door usually gets the occupant's attention. When water drips from that drain, the occupants should get the primary drain cleaned. In other cases, there is a problem installing a secondary drain line. In this case, a water float switch is installed in the auxiliary drain pan. When water starts to collect in the pan the float raises and opens the control circuit turning the system off. This float switch can be placed in the pan that has the drain line piped to it also, in the event its drain becomes inoperable.

CHAPTER 18

Optional Controls

Chapter 9 discussed some of the many safety controls that the service technician will find in residential and light commercial equipment. Many air conditioning units are made without safety controls installed within them. The reason for this is to reduce their cost. This chapter explains those you should be aware of; some of these controls might be installed in a unit at a later date if they are needed.

FIRE STAT

This control might be required by law in certain regions. The control actuates when a predetermined amount of heat is present. This control immediately shuts down the entire system when excessive heat is sensed. Its operation is based upon a bimetal reaction. This control is equipped with a manual reset. In some commercial applications, this switch can be wired through a fire alarm system that shuts down any other air conditioning systems in the same building while turning in the alarm to the local fire department. Figure 18-1 shows a typical fire stat.

SMOKE DETECTOR

Many homes today have smoke detectors that sound an audible signal to alert its occupants. The smoke detector in the air conditioning system stops the operation of the machine. The reason is to prevent the spread of smoke throughout the structure. This

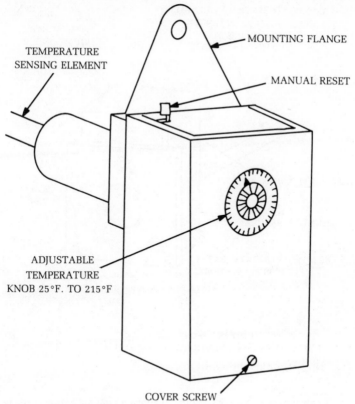

TEMPERATURE
SENSING ELEMENT

MOUNTING FLANGE

MANUAL RESET

ADJUSTABLE
TEMPERATURE
KNOB 25°F. TO 215°F

COVER SCREW

Fig. 18-1. Fire stat.

switch also has a manual reset. Smoke detectors can also be installed on ventilation fans to stop their operation when smoke is detected. Figure 18-2 shows this type of smoke detector.

SAIL SWITCH OR PRESSURE SWITCH

This type of switch is used to monitor the pressure differential on both sides of the evaporator coil. The sail switch has a small plastic sail that is counterweighted with a small spring to keep the sail off center. When this unit is placed in the supply air stream, air velocity keeps the sail on center. If the velocity drops, the spring pulls the sail off center and opens the switch. This condition can happen if the filter becomes clogged, the fan isn't moving enough air, or the coil is dirty. The control can be wired to set off an alarm system to alert the occupants, or it can be wired to turn off the system.

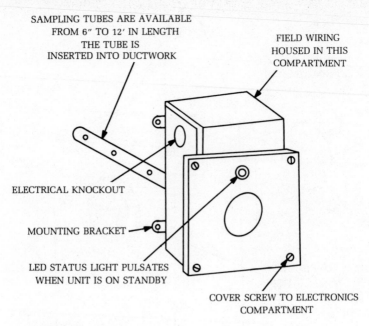

SAMPLING TUBES ARE AVAILABLE
FROM 6″ TO 12′ IN LENGTH
THE TUBE IS
INSERTED INTO DUCTWORK

FIELD WIRING
HOUSED IN THIS
COMPARTMENT

ELECTRICAL KNOCKOUT

MOUNTING BRACKET

LED STATUS LIGHT PULSATES
WHEN UNIT IS ON STANDBY

COVER SCREW TO ELECTRONICS
COMPARTMENT

NOTE
THIS TYPE OF UNIT IS USED IN AIR
CONDITIONING AND VENTILATING SYSTEMS

Fig. 18-2. Smoke detector.

A pressure switch works with a diaphragm having a tube sensing pressure on each side of the coil. This switch can be wired just as the sail switch is. Some residential units are equipped with a switch like this which is wired to an indicator light on the thermostat. The device helps avoid refrigerant slugging and decreases the running time of the compressor by telling the owner there is a restriction. Figure 18-3 illustrates these components.

OIL PRESSURE FAILURE CONTROL

This switch is used on semi-hermetic compressors to protect them in the event there is an oil pressure problem causing the lack of lubrication of the reciprocal compressor. The operation of this control is relatively simple, yet many technicians become confused when working with the control. The opfc is a pressure actuated control, using two opposing pressure diaphragms. The capillary tube of one diaphragm is connected to the oil pump port. It senses the actual oil pressure being produced by the pump. The other capillary

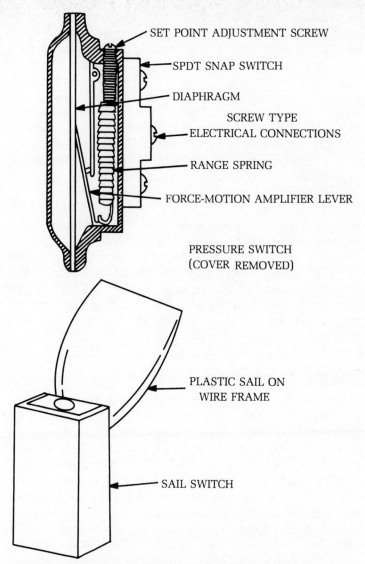

SET POINT ADJUSTMENT SCREW

SPDT SNAP SWITCH

DIAPHRAGM

SCREW TYPE
ELECTRICAL CONNECTIONS

RANGE SPRING

FORCE-MOTION AMPLIFIER LEVER

PRESSURE SWITCH
(COVER REMOVED)

PLASTIC SAIL ON
WIRE FRAME

SAIL SWITCH

Fig. 18-3. Sail switch and pressure-diaphragm type of switch.

tube from the other diaphragm is connected to the low side (suction) of the system. For example, the oil pump has an output pressure of 30 psi. It is an R-22 system with a 65-pound back pressure. A spring will apply pressure on the oil pump diaphragm of the control. The spring is adjusted to a 35-psi pressure causing equal pressure on each side of the control. With a usual 10-pound differential the control switch will not close unless there is more than a 10-pound deviation on either side of the control. In the event that there is a pressure drop of more than 10-psi a switch, that supplies energy to a resistive heater, closes. The resistive heater causes the flexing of a bimetal switch that opens the control circuit. In doing so the entire condensing unit shuts down. This is a manual re-set control. Most of these controls can be wired for 120 or 240 Vac. There are also different resistances for the heaters. The reason for this is to allow a time lag before the control circuit opens. Their standard switches are made to open the main control circuit in 60, 90, or 120 seconds from the time the heater becomes energized.

The reason for the time delay is to give the system time to stabilize its pressure. If it is a pump-down system, it might take the compressor 80 seconds to pump down. The control has to allow for this time or the service technician would be constantly called to push the reset button. When a TEX valve (temperature expansion) is in a system, there is a time on an initial start when the valve will hunt, thus fluctuating the pressure. The time delay allows the TEX valve a short time to hunt and level out. The principle of this control is a simple one, and it is illustrated in Fig. 18-4.

SOLENOID VALVE

The solenoid valve is an electrically-operated stop valve. You are already familiar with manually-operated stop valves such as the gate, globe, and king valves. All of these valves need a manual force to operate them. A solenoid valve is operated by electromagnetic energy. These valves are designed to be used with liquids or gases.

The coils that operate the pilot valve are made in many voltages. Those most commonly used are 24 Vdc, 120 or 240 Vac. Although the valves are designed to operate in any position of installation, it is suggested that they be mounted as to have a horizontal flow. The valves are also designed to be directional; this is designated with a flow arrow on the valve body. Figure 18-5 shows this feature.

A small pilot valve is operated with the use of the magnetic force exerted on it to lift it from its seat. A force is then exerted upon the valve diaphragm by the pressure of whatever the medium is flowing through the valve. Solenoid valves are designed to be installed either by flaring them into the circuit or soldering them into the circuit. If you are soldering a solenoid into a system, remember to disassemble the valve before applying heat to it. The body should

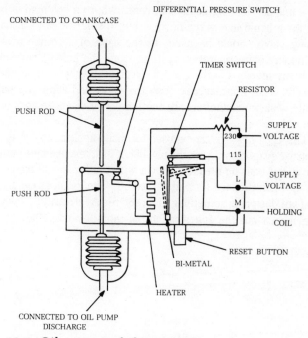

Fig. 18-4. Oil pressure failure switch.

also have a wet rag or other type of heat sink to absorb excessive heat as not to warp the valve body causing inconsistent of possibly no operation.

Another important factor about solenoid valves that I want to point out to you at this time is the possibility of installing a solenoid

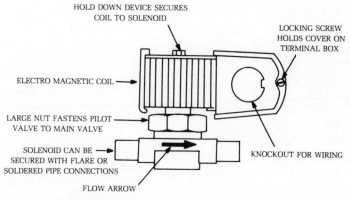

Fig. 18-5. Solenoid valve.

171

with a manual opening device. In cases where a solenoid is being used for a liquid line pump-down valve, a solenoid with a manual opening valve should be used. In the case of a commercial air conditioning or refrigeration system, when the holding coil fails or a problem develops in the control circuit, the valve can be opened manually. This will manually open the valve and allow the unit to operate while you either secure another coil or diagnose and correct the problem.

This type of valve can be used for many applications. It is used to defrost evaporators with the use of hot gas, it is used for pumping down a unit causing it to shut down with the low-pressure control switch, it can be used for water turn off. The list goes on and on. If you are designing something that requires an electrically-operated valve, check out the specifications of the solenoid valve.

CHAPTER 19

Piping

The thing that I enjoy the most being in the air conditioning and refrigeration industry is the diversification. When working in HVAC (heating-ventilation-air conditioning) you retain your interest by meeting new challenges on each job. Your expertise will include your capability to do some piping. Piping will include high pressure, medium pressure, and low pressure piping. In our industry, the right type of piping must be used for the right job.

COPPER

Copper is the most used material in our industry. It is manufactured in many forms. Hard drawn copper is rigid and is identified by inside diameter and letters. The letters designate the purpose for which the pipe or tubing can be used. Rather than having to place the thickness of the wall in the copper, the letters inform the technician about the uses of the pipe. The most popular type used for pressure piping is type L. ACR and L are used for air conditioning and refrigeration. They are dehydrated upon their manufacture and sealed on both ends to keep moisture and other contaminants from entering the pipe. Types L and K are usually used for low pressure and drain lines. Hard drawn copper comes in 20-foot lengths.

Soft copper tubing is also used extensively. The rolls manufactured for ACR (air conditioning and refrigeration) use are dehydrated and sealed as is the hard drawn. Advantages of soft copper are the ability to bend, and its length. It is sold in 50-foot rolls. This allows a 50-foot run of copper without having the need of a joint, a potential leak. Applications where piping will be enclosed in a wall or floor, are the perfect places to use the soft copper. Joints that are placed in inaccessible locations are usually the ones that leak the most. If you can make a piping run in one piece, do it. In certain commercial and residential construction, a raceway is placed beneath a concrete flooring. Hard drawn copper is difficult to pull through the raceway and there has to be a joint every 20 feet of run. The soft type will pull easier, and you will only have half as many joints.

When copper is installed above drop ceilings, across attics, or up interior of walls, support brackets should be used. If a pipe is to be used for a suction line, it should be insulated to prevent condensation. This can cause a lot of damage at a later date. Insulation should be used, and all insulation joints should be sealed properly. Insulation tubing is made of foam rubber and is sized according to its inside diameter and wall thickness. Insulation is supplied in lengths of 10 feet, so you can see that a long run of piping requires many joints in the insulation. They must be sealed to keep warm, moist air from contacting the cool suction line.

Good practice in piping of the suction line is to pitch the line down slightly in the direction of the compressor. This causes any oil in the pipe to return to the compressor. It is good piping practice to use traps in the suction line every 10 feet of rise. See Fig. 19-1 for an illustration of traps. This is done only when the compressor is mounted higher than the evaporator. The trapping keeps oil from loading the evaporator and lets the oil get back to the compressor crankcase faster. The traps hold oil when the unit shuts down.

Soft copper can be jointed with either solder fittings, flaring fittings, or compression fittings. Hard drawn can only be soldered. When cutting copper, never use a saw, use a tubing cutter only. Copper filings have a tendency to create havoc in a system. Copper filings can penetrate the winding insulation and cause a compressor to ground out electrically. Use a tubing cutter and a reamer to remove the internal ridge left in the copper.

PVC

Plastic piping is used extensively in our industry. It is primarily used for drain lines and low pressure piping. The PVC pipe also comes manufactured with different thickness of pipe wall. Schedule C is the most used in air conditioning and refrigeration due to the thick walls of its construction. Thin wall pipes are used in other applications such as lawn sprinkler systems.

174

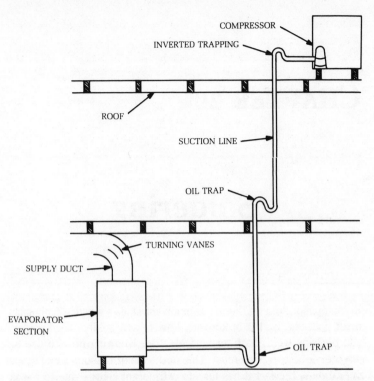

COMPRESSOR

INVERTED TRAPPING

ROOF

SUCTION LINE

OIL TRAP

TURNING VANES

SUPPLY DUCT

EVAPORATOR
SECTION

OIL TRAP

Fig. 19-1. Oil trap in a vertical riser when the condenser unit is located on a roof over two stories high.

PVC drain lines must be supported at close intervals. If pipe is located in hot environment such as the attic or across a rooftop in the sun, it will soften and sag if not supported properly. PVC piping is also now being used for water piping of a water-cooled unit. The PVC is seen on water towers and piped in the walls of condominiums to supply water from tower to each apartment. The PVC industry has grown to a point that they not only manufacture fittings for PVC pipe, but have manufactured and developed a full line of water valves including throttling valves. When the proper installation procedures are used, PVC is a long lasting piping medium.

GALVANIZED

This type of piping is rarely used on residential or light commercial application. It requires heavy bulky equipment to cut and thread pipe. Galvanized, black iron, and steel pipes are used primarily on large commercial applications.

175

CHAPTER 20

Soldering

In my opinion, one of the most important skills a service technician must possess, is that of knowing how to soft solder, hard solder, and perhaps weld, both gas and electric. Knowing how to use an oxyacetylene torch is a must. One of the major reasons a unit stops operation is the fact it has lost its refrigerant charge due to a leak that has developed. It is an absolute necessity that you locate the leak and make a professional repair. If this is not done, re-charging the unit and leaving, telling the customer the unit has been repaired, will only lead to a dissatisfaction between you and the customer. It can also mean a loss of money due to the fact you can't charge the customer for something that should have been fixed the first time.

In soldering, the most important tool is the oxyacetylene torch. It can supply a wide range of heat to solder different metals. As a service technician, I've found that the MC size tanks are the most convenient. They are easy to handle in close areas and are not too heavy carrying to a roof on a ladder. For service work, these small bottles allow great mobility. In construction or installation, the larger tanks should be used. The MC tanks do not hold enough gas to allow many hours of uninterrupted soldering. Larger tanks equipped with longer hoses, allows for longer periods of soldering, without having to haul the heavy weight of the tanks. Using straight acetylene or propane can be used for low temperature soft solder joints. With hard solder, temperature must be higher to attain a good joint. Solder must be selected for the proper job you are doing.

50-50 SOLDER

This type is the least expensive solder. In all soldering, the pipe must be prepared by cleaning its surface for bonding. This is done with many different types of abrasives. I prefer sanding cloth. The materials used in 50-50 solder are tin and lead in equal parts. Paste flux can be used with 50-50 solder to clean the mating surfaces. The flux contains an acid, such as muriatic (hydrochloric). The joint should be wiped with a damp rag when it is complete. If flux is left on the copper wires in the compressor. On certain older installations of an melting temperature. It is used mostly on water lines. Condensate drain lines made of copper are soldered with this type of solder. This type of solder might be used on one joint in the condenser coil of a unit to act as a pressure relief device. Exercise extreme caution when using soldering paste flux so as not to get it inside the refrigerant piping. As it attacks the copper piping, it can attack the copper wires in the compressor. On certain older installations of an R-12 system, 50-50 solder was used on the low side of the system. This was a dangerous practice due to the lead decomposing and causing leaks and also joints blowing out when high pressures were achieved due to a malfunction in the unit.

95-5 SOLDER

This type of solder can withstand higher pressures than 50-50 and requires low heat to apply. The solder is made from 95% tin and 5% antimony. Without the possibility of lead deterioration, this solder can be used on refrigerant piping. This solder requires a flux to be used with it. One of the biggest problems found using this solder is it becomes very fluid when heated. If a joint needs to be filled, it can be difficult using this type of solder.

STAY BRITE

This solder has a low melting point like 95-5, yet it has a heavier flow characteristic. It can fill a joint. The price of this solder is a little more than 95-5. Test pressures for this solder are well within the limits to be used on all refrigerant lines that use halogen refrigerants.

SIL-FOS

This type is the most popular of hard solders. It contains a small amount of silver in it that gives it strength. Price of this solder is much higher than the price of those already discussed, in my opinion yet it is the best. It does require a lot of heat to make a good fusion at the joint. The copper has to be heated to a point where it almost turns red. A joint made with this type solder is actually stronger than the copper itself. With a silver solder flux this solder can be used to create a joint between two dissimilar metals such as steel and copper. This is found to be a situation very common in replacing in-line driers and compressors with steel ports. It can be used to join copper to brass as when you are soldering a brass valve body to copper. Remember to use wet rags or some other type of heat sink to help prevent the distortion of the body when such high temperatures are used in soldering. Although this type of solder contains only a small percentage of silver, it is fairly expensive.

SILVER SOLDER

This in our industry is the top of the line in solder. Its higher silver content makes it flow and fill very easily. A special flux is used and the material to be soldered has to be very clean before a joint is attempted. The flux used with silver solder is a white paste that sometimes dries and becomes hard in the jar. It is water soluble and can be made ready for use by adding a little water and mixing thoroughly.

I want to place a warning at this point. Whenever you are soldering, make sure there is adequate ventilation in the work area. Many of these solders release toxic fumes when used. Not only does the solder emit fumes, but the refrigerant vapors can. For this reason, ventilate even if it means having to use a floor fan to move the air.

Several years ago, in order to hold costs of air conditioning down when copper prices escalated, other materials were sought to be used in the manufacture of evaporator coils and condenser coils. The answer was aluminum. Some feel it was a good breakthrough in the industry while others thought it was another source of trouble. Although the heat transfer factor was very good in comparison with copper, the expansion and contraction factor was not. Copper seemed more flexible when going from hot to cold and vice versa. After a couple of seasons of operation with aluminum coils, some manufacturers found their products in trouble. Repair kits were made available along with other repair products such as epoxy glues. The service technician found themselves in a bind. Finally an aluminum solder was developed that worked very well. It takes a little skill to use it. Very little heat is used, and when applied to the area being repaired, it has to be moving constantly. This solder has its own special flux. A technician should practice with this solder on an old aluminum coil to become familiar with the way it flows.

Whatever type of soldering you do, the first and most important step is preparing the work for solder by cleaning the pipe and fittings. Some say it is not necessary to clean new copper. This is not true, for there is always a very fine film of oxidation on copper. It is possible to make a joint without cleaning; however, it will not be as strong and permanent as a clean joint. Cut squarely with a tubing cutter. Clean interior ridges left by the tubing cutter. Whatever you use to clean the copper surfaces, be careful not to get any of the grit inside of the tubing. This will lead to problems. In some cases such as the ports of some new compressors, steel is used that has a copper tinning on it. Sand the ports gently so as not to remove the copper. When soldering an assembly such as a solenoid valve, or a TEX valve, disassemble it before soldering in order not to damage any of the internal parts with heat. In cases such as with the reversing valve of the heat pump, you might have to wrap the valve in wet rags, or place the valve in water while placing copper stubs to the ports of the valve. Then the valve can be soldered into the unit without having to place the heat near the valve body itself. Always keep the flame pointed away from the work. One more thing, a piece of sheet metal should be kept. A piece about 12 inches square works out well. There will be times when a joint is near an insulated panel or close to a piece of timber, the sheet metal is used as a heat shield between the work and the inflammable material. A fire extinguisher should be right at hand when soldering, especially in those tight places in the crawl attic. Some might laugh, you won't the first time you witness a flash dust fire in an attic. When soldering outside the structure on some of the piping of the condensing unit, wet down the area with a water hose. At least have the hose hooked up and next to you. Solder slag in dry grass or an oil-soaked cabinet wastes no time in creating a fire. I said in the very beginning of the book, "the sign of a good technician is one that thinks out the job before jumping into it."

Due to excessive heat that is applied to the copper during the hard soldering process, carbon flaking takes place on the inside and the outside of the pipe. This flaking can cause many problems in the system, such as restricting the expansion valve orifice or preventing a solenoid valve from closing completely. The list of problems is endless. To prevent this from happening, an inert gas is used to displace the air (oxygen) within the pipe. Without oxygen the flaking doesn't occur. Nitrogen is used for this purpose. This gas is harmless and non-inflammable or explosive. It is also inexpensive to use. It is introduced into the system through a pressure regulator and the pressure is adjusted just a couple of pounds above atmospheric pressure in order to displace the air. Soldering of the joints can then be done without carbon flaking taking place. I strongly urge the use of nitrogen when soldering.

I can't impress the importance of never trying to use nitrogen in a system without the use of the proper regulator on the nitrogen tank. The pressure of this inert gas is extremely high and can rupture pipes, sight glasses, and blow-out terminals. You are taking an unnecessary risk when you don't use a regulator with nitrogen. It is like playing with a hand grenade and waiting for it to go off.

In closing this chapter, I would like to give you one more suggestion. When piping and soldering in an area where there is a high concentration of vibration, use some type of vibration eliminator. Figure 20-1 illustrates a vibration eliminator. If one like this is not available, vibration loops can be made from the copper stock. In Fig. 20-2 a loop is shown that is used when soft copper is used. In Fig. 20-3 a loop is shown when hard drawn copper is used. Regardless which method used, something should be used to absorb the vibration. If nothing is used, the pipe will eventually crack. There will also be continual cracking in the line with the vibration until the eliminator is installed.

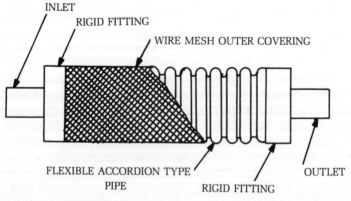

INLET

RIGID FITTING

WIRE MESH OUTER COVERING

FLEXIBLE ACCORDION TYPE PIPE

RIGID FITTING

OUTLET

Fig. 20-1. Vibration eliminator.

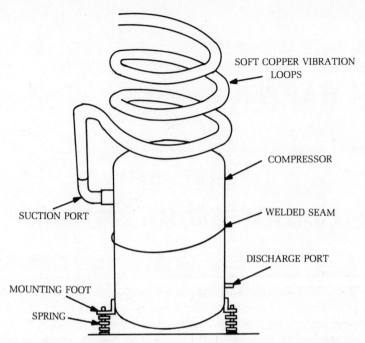

Fig. 20-2. Vibration loop when soft copper is being used.

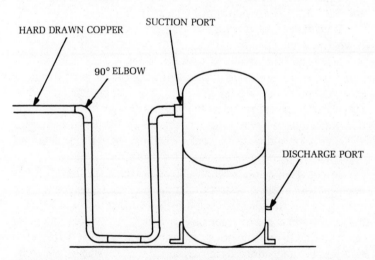

Fig. 20-3. Vibration loop when hard-drawn copper is being used.

CHAPTER 21

Flare and Compression Fittings

There are two other methods that are used in the industry to join pipe. Flaring is one method, and it can be used on piping that will carry liquid or gas. The other method is compression fittings. It too can be used for both liquid and gas piping. In fact, the compression method of piping is used on small plastic piping such as that used in ice machines. The compression fittings are shown in Fig. 12-1, and flare fittings are shown in Fig. 12-2.

Remember, you can't always use an open flame to join pipes. If working on heating fuel lines such as gas or oil you can see that an open flame is impossible to use. If it is a new installation and the pipes are clean, soldering is safe; however, in a repair, flaring or compression fittings are the answer. Care must be executed when preparing the piping in this method just as if you were going to solder. Remember a few minutes extra spent on making joints in pipes will save a lot of time later. Leaks will be avoided if a professional approach is taken preparing pipes for jointing.

In Fig. 12-3 a side view of a piece of copper is shown. It has been cut with a tubing cutter. Under close examination you will see a ridge on the cut edge that is left by the cutting wheel. This must be removed in order to accomplish a tight joint. By holding a piece of sanding cloth or some other similar material on a flat surface, the ridge can be dragged across the flat surface and is removed squarely. This squareness in both types of jointing is essential for a good joint. If ignored and left on the edge of the tubing, the tubing would not

182

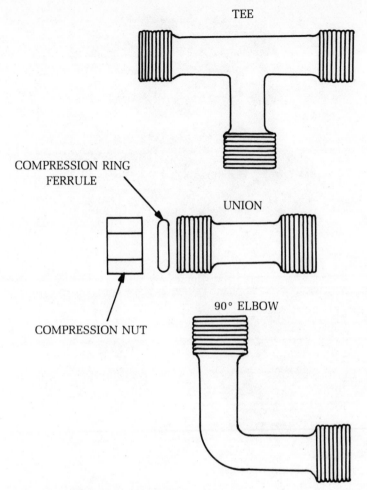

TEE

COMPRESSION RING
FERRULE

UNION

COMPRESSION NUT

90° ELBOW

Fig. 21-1. Compression fittings.

mate properly inside the fitting. With the ridge removed, another ridge on the inside of the tubing must be removed. This is quickly done by reaming. The tubing is now ready for joining.

Flare nuts and fittings are made of brass. The whole principle of flaring is placing copper between two brass mating surfaces. The copper being a softer metal will compress making a seal. With this in mind, when flaring, do not compress copper in flaring block excessively. A good flare is shown in Fig. 21-4. Notice how the copper flare mates exactly to the brass fitting mating surface. If the

183

Fig. 21-2. Flare fittings.

flare is too shallow, eventually a leak or separation will occur between the copper and brass. A shallow flare makes a weak joint. Over flaring doesn't present a problem due to the fact that the flare nut will not pass over the flare to mate with the fitting.

If you are involved with light commercial refrigeration equipment, it might be advisable to purchase a set of flare nut wrenches. These are shown in Fig. 21-5. This type of wrench helps preserve knuckles and flare nuts. An adjustable wrench is commonly used; however, these have a tendency to round the edges of the flare nut and cause the wrench to slip. The flare wrench holds the flare nut at each corner surface. In refrigeration work, the suction line is constantly wet or frosting; when the compressor stops operating, rust and corrosion

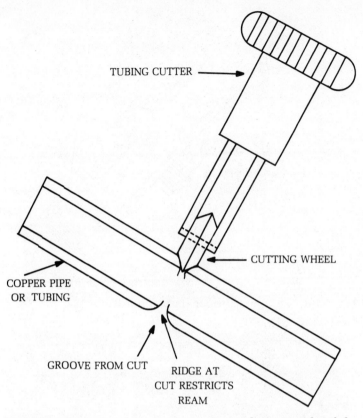

TUBING CUTTER

CUTTING WHEEL

COPPER PIPE
OR TUBING

GROOVE FROM CUT

RIDGE AT
CUT RESTRICTS
REAM

Fig. 21-3. Copper tubing cutter causes ridging inside of the tubing.

can form between the flare nut and the copper pipe. A solvent and/or heat might have to be applied in order to loosen the flare nut.

You must remember that a pressure tight joint can't be accomplished when two hard metals such as brass are put together without a soft metal acting as a gasket between them. This applies to both steel and brass. Soft seats as shown in Fig. 21-6 are used at ¼-inch valve ports. Caps are placed on these ports that can be made of brass, aluminum, or plastic. When a unit is new, the service valves and ports might not leak; but when they age a bit, they are a primary source of refrigerant leaks. A soft seat corrects this problem. Another way to approach the problem is with a flare nut using a bonnet or a pig-tail. These are shown in Fig. 21-7. Another location where soft seats should be used is with in-line filter driers with flare connections. An in-line sight glass is another example. These

185

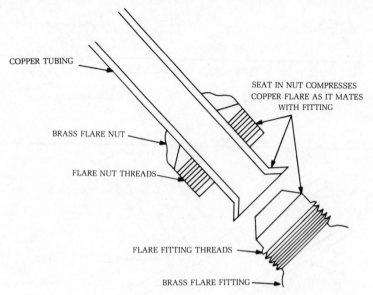

COPPER TUBING

SEAT IN NUT COMPRESSES
COPPER FLARE AS IT MATES
WITH FITTING

BRASS FLARE NUT

FLARE NUT THREADS

FLARE FITTING THREADS

BRASS FLARE FITTING

Fig. 21-4. Making a proper flare connection.

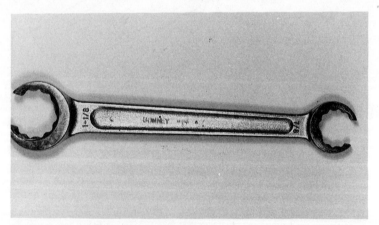

Fig. 21-5. Flare nut wrenches.

components are usually made of brass or steel. Remember, when
two hard metals are to be placed together with the use of flaring,
a soft seat should be used between the mating surfaces. With some
service valves as with a king valve, a soft copper circular gasket
should be used when replacing the valve stem cover nut. Around
the base of the valve stem, outward from the gland nut, is a machined

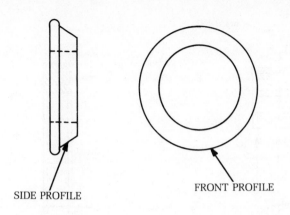

SIDE PROFILE   FRONT PROFILE

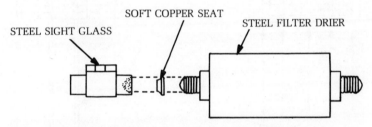

STEEL SIGHT GLASS   SOFT COPPER SEAT   STEEL FILTER DRIER

Fig. 21-6. Soft copper seats and an example of use.

surface to mate with a surface inside the cap nut. The copper gasket acts the same way as the soft seat. In Fig. 21-8 this type of cap nut is shown. This kind of service valve arrangement is found on equipment from fractional horsepower compressors through large horsepower units.

Compression fittings do not require special tools like the flaring block, which serves like a kind of vise. The copper must be cut square and cleaned as described earlier. The compression nut is placed on the copper pipe followed by a brass ring. This ring, called a ferrule, looks like a miniature wedding band. The end of the copper that is to be jointed is placed into the compression fitting. The copper tubing is held against the seat of the fitting, and the ferrule ring is slid against the fitting. The nut is then brought to the fittings threads and is tightened on to the fitting. Hold copper firmly against the seat of the fitting until the tightening is completed. The parts and procedure are illustrated in Fig. 21-9.

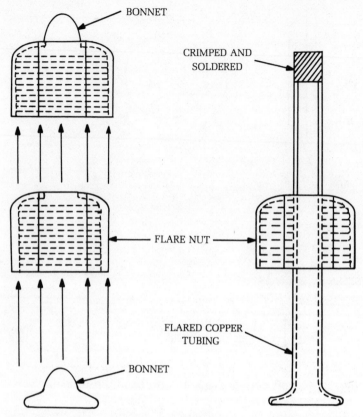

Fig. 21-7. Copper bonnet and pigtail.

Compression fittings work on a different principle than the flared fitting. With the flared fitting, an angled surface is formed at the end of the tubing, and this surface acts as a soft seat mating the two materials together. The compression fitting has a tapered receptacle where the copper or plastic is inserted and seated. The ferrule ring when pushed by the nut begins to compress as it is pushed into the receptacle. The ferrule ring actually compresses into the copper. It does make a good seal and is also used on plastic and aluminum piping.

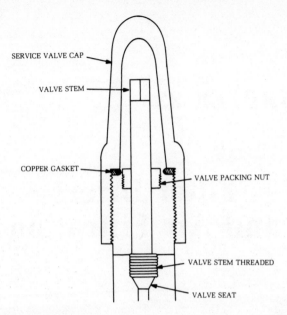

Fig. 21-8. Service valve with cap nut used to seal valve from leaking refrigerant from the valve stem.

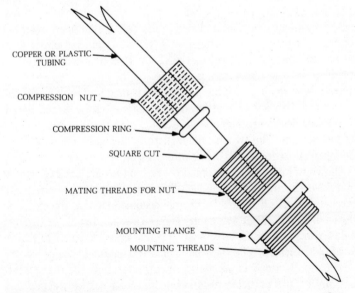

Fig. 21-9. Exploded view of the procedure to make a compression fitting joint.

CHAPTER 22

Filter Driers
and Air Filtration

In the first part of this chapter, I want to discuss filter driers that are used in both air conditioning and refrigeration systems. To begin with, look at the illustration in Fig. 22-1. This shows how a typical filter drier is put together.

The filter drier that you will probably be most familiar with is called an in-line drier. It is placed in the refrigerant lines by either flaring or soldering. Look again at Fig. 22-1, this particular filter drier has a direction of refrigerant flow. It is usually indicated on the body with an arrow, or it is stamped on the ends, inlet, or outlet. This is very important. If the component is mounted backwards, the refrigerant flow will be restricted. Pay strict attention to this detail.

Filter driers are made for the liquid side and the suction side of the system. The two are constructed differently and should not be installed to do the job of the other. First, a liquid line filter drier is shown in Fig. 22-1. The liquid refrigerant enters the component and flows through a metal mesh. This mesh traps large particles of foreign materials that would ordinarily become lodged in the metering device if no filter drier was used. The refrigerant then flows through a desiccant. Desiccants are compounds that have the ability to absorb and hold moisture. The ones used in air conditioning are called dry desiccants. Not only is the moisture extracted from the refrigerant, but any minute particles missed by the mesh will be trapped and held in the desiccant. A new type of liquid line filter drier was introduced with the advent of the heat pump. It is an in-line filter

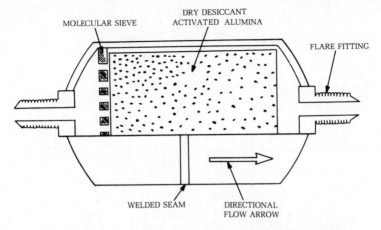

MOLECULAR SIEVE
DRY DESICCANT
ACTIVATED ALUMINA
FLARE FITTING
WELDED SEAM
DIRECTIONAL
FLOW ARROW

Fig. 22-1. Cut-away view of a dry desiccant filter drier.

drier that filters and dries when the refrigerant flows in either direction. It is called a bi-flow or bi-directional filter drier. When installing liquid line driers, try to install them in a vertical plane, if possible. This orientation will provide better filtration of the refrigerant. Figure 22-2 shows filter driers mounted in both positions in a system that is slightly undercharged. It is easy to see that more surface of the dry desiccant is being used in the vertically mounted unit.

Liquid line filter driers and suction line driers are sized according to their cubic inch desiccant area, type of connection for piping, and pipe size of inlet and outlet ports. The first two digits designate the cubic inch capacity of the unit. The third digit designates the pipe size fittings in eighths of an inch. The letter S at the end designates soldered fittings. If no S is present, the unit accommodates flare fittings. Example: the number on a filter drier 083-S would indicate that this was an eight cubic inch with ⅜-inch solder fittings liquid line drier. Sizes of this type of filter drier range from three inches to 75 inches. These would be more than sufficient for anything found in light commercial and residential equipment. For the sake of information a core shell type of filtration system is available for large equipment. The shell core type is available up to 400 inches of desiccant area.

Suction line filter driers are constructed somewhat differently than the liquid line filter driers. They are sized the same way, and their bodies look almost the same except for a ¼-inch flare port on the inlet side of the unit. This is a pressure test port. You can

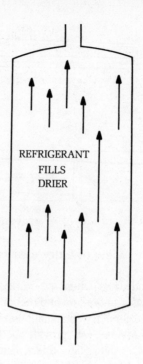

REFRIGERANT
FILLS
DRIER

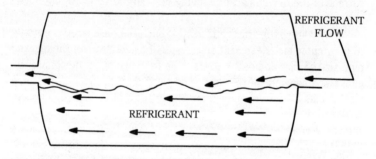

REFRIGERANT
FLOW

REFRIGERANT

Fig. 22-2. Filters mounted in vertical and horizontal planes.
The filter drier mounted in the vertical plane operates more
efficiently when the refrigerant charge is low.

measure any pressure drop by attaching a compound (low pressure)
gauge between the port on the drier and the suction side of the
compressor. These suction line filter driers are designed not to have
more than a two to three pound pressure drop. If the drop exceeds
that, the filter is restricted and should be replaced.

The shell-and-core type filter drier can be used on either refrigerant line. The smallest, available size of this unit varies with the manufacturer. One makes a steel body shell with a minimum size of 48 cubic square inches. Another manufacturer makes a brass bodied unit that does have a slightly smaller core capacity. The main advantage of these units is the ability to select the core type to be installed. They come in standard core, high capacity core, and cleanout core. In the event of a compressor burn out (grounded), when acid tests show positive, these cores are used and changed after a couple of days of operation. The procedure of changing cores is repeated until the system is free from acid or any other contaminants. In Fig. 22-3 a shell-and-core type filter drier is shown.

In this book, the proper way is being taught. In the field when dealing with a customer, price becomes very important. Sometimes it is the determining factor as to how a piece of equipment is repaired. Yes, you will encounter many that only want a band-aid, rubber-band, baling wire type repair, but you should know how the prescribed methods are used. If filter driers are used on the system, they should be piped with a by-pass fitted with valves. Figure 22-4 shows the method. This type of installation is used on large commercial units for two reasons. The first is that a technician can quickly re-start a unit that shut down due to restricted filter driers by removing the filter driers with the valves. Second, if the system has a large refrigerant capacity, some of it might be lost when changing the filter driers. This can be costly on large commercial units.

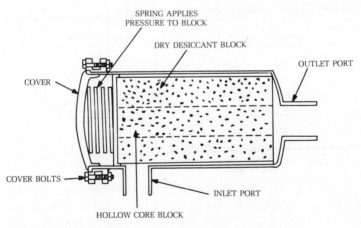

Fig. 22-3. Cut-away view of a shell-and-core drier.

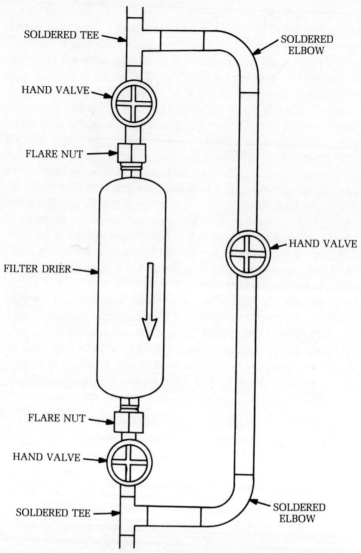

Fig. 22-4. One method to install hand valves that enables the filter drier to be changed without losing refrigerant charge.

In most small units, the metering device is a capillary tube. Air conditioning and refrigeration systems use this type of metering system. A very fine strainer is used to trap particles from entering the small orifice of the capillary tube. The very fine mesh is the only thing in the body of the filter, sometimes called strainer. In Fig. 22-5

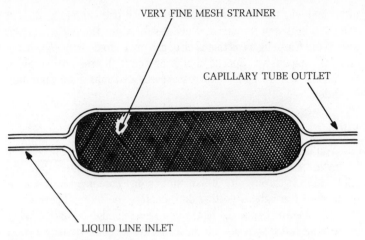

VERY FINE MESH STRAINER

CAPILLARY TUBE OUTLET

LIQUID LINE INLET

Fig. 22-5. Filter strainer.

a filter strainer is shown. This unit is immediately in front of the capillary tube.

The ideal placement of any liquid line filter drier or strainer is directly before the metering device. It can trap any particles that might be in the liquid line between the condensing unit and the evaporator section. This holds true to the straight cool systems that use the directional-flow type filter drier. In a heat pump, it doesn't matter where in the liquid line the bi-flow filter drier is installed.

AIR FILTRATION

Air filters are the least expensive component in an air conditioning system and perhaps one of the most important, yet it is overlooked by many. When the return air filter becomes restricted, air can no longer be circulated through the evaporator coil. First, this condition decreases the efficiency of the unit and increases its operating cost. Second, damage to the compressor will be inevitable. The only way refrigerant can evaporate in the evaporator is to absorb heat from the air passing over it. With the air being restricted by a dirty filter the liquid passes through the evaporator and goes back to the compressor in the liquid state. Liquid cannot be compressed with this type of equipment, so eventually, the compressor is destroyed. Usually the damage resulting from this condition includes, broken pistons and connecting rods. When a compressor sounds noisy, you should check for a slugging condition. The liquid slugs the pistons, thus a slugging condition. If a units filter is not changed

195

often enough, the evaporator coil itself will become clogged. Dirt will fill in between the fins and eventually cover the entire surface of the coil. Cleaning a coil that is dirty can prove to be time consuming and very expensive. Special chemicals and high pressure spraying equipment will be needed to restore the coil into good operating condition.

Filter Changing

You'll be asked how often should a filter be changed. The answer is simple, whenever it becomes dirty. If you hold a filter up to the light and look through it, the amount of dirt contained in it will be obvious. If the light is getting dim, change the filter. There can't be a set time to change the filter due to everybody's situation being different. A home with several children and a few pets running around will cause a filter to load faster than a filter located in a home of an elderly couple. Another rule of thumb is about once a month the filter might need changing.

Types of Filters

There are many types of filters on the market. Naturally the more efficient a filter is, the higher the cost. A good cheap filter will allow dirt to pass through it and restrict the evaporator coil. Unfortunately too many people are using that type of filter. Many units are shipped from the factory with a so-called, permanent, hog-hair filter. It is neither. This particular type of filter medium should be washed and sprayed with a filter coating that bonds dust particles to it. It has varying sized voids in the material and allows air to flow unfiltered to the evaporator coil.

Second, the most popular filtering medium is the spun glass fiber filter contained in a cardboard frame. It is usually one-inch thick for residential application. This type of filter can be purchased in two-inch thickness for commercial units. This filter is sold in most stores and building supply outlets. They are inexpensive, but they too let a lot of dirt get through. In some situations, a filter is an odd size and the owner has to have a filter built specially for it. An aluminum frame filter is the most popular. This type can have a choice of media placed in it. Shredded aluminum or steel fill are considered permanent type filters. The shredded mesh medium can be washed with water, and then must be dried and sprayed with a filter coat. The aluminum frame can be filled with a dacron medium that works very well. The dacron medium pad is discarded when dirty, and a new pad is installed in the original aluminum frame. This dacron medium is very effective. Static filters made of a material similar to the fishing line monofiliment are used in lithographic companies to trap paper dust. This is about the finest of the mechanical type filter. Some window units use a piece of thin sponge. These too are not very effective. Remember, on some of the permanent type filters, a filter coat must be sprayed on the medium after it has been cleaned.

In some commercial applications, such as a bowling alley, not only is dust a problem, smoke and odors present special attention. The primary filter as described in the last paragraph is used for the entrapment of large particles of dust and dirt. As a second stage filtration, charcoal can be used to remove some smoke and odors.

Filter Efficiency

Filter efficiency is rated by the percentage of foreign particles trapped by the filter. Flat, spun fiberglass, disposable filters have a rating of about 15 to 20 percent. The pleated filter has a 30 percent efficiency. These filters, due to the pleating, have up to seven times the filtering surface than a flat filter has. This allows them to stay in service for longer periods of time. Bag filters, used on large evaporators, are deep pleated and have an efficiency rating as high as 55 to 95 percent. These filters could be used in laboratories, helping those with hay fever or allergies. Oil paintings can be protected by this type of filter. It is used in what the industry calls, "the clean room."

Magnamedia or Absolute Filters

The efficiency of this filter will rate from 95 to 99.9 percent. This is a honeycomb-type filter with aluminum or cardboard separators and a glass fiber medium. The filter can withstand a temperature as high as 800 degrees F., and the humidity can be as high as 100 percent in certain cases.

The clean magnamedia or absolute air filters will have a manometer reading of ¾-inch water column. This filter should be changed when the manometer reads 1.5-inches water column.

Electronic Filter

This is also known as the electrostatic precipitator. Many evaporators use this type of intermediate filter. It consists of two parts, grid wires and plates. The plates are mounted vertically and spaced about one-half inch apart. The grid wires are approximately two inches ahead of the plates and between the plates. They are mounted in a vertical position, and the grid wires themselves are suspended by high voltage insulators. Plates and grids are made from either aluminum or stainless steel.

The electronic filter operates on high voltage, over 10,000 volts dc, placed on all the grid wires. Plates are usually at ground potential. When the air stream passes the grid wires, all the particles become positively charged. With the particles having a positive charge, they adhere to the plates that are negatively charged. On some units, the entire air conditioning system will shut down (stop running), and an indicator light and/or alarm will activate telling the owner the filter needs cleaning. The above-mentioned unit requires manual disassembly and cleaning. There are units that activate their own cleaning systems. Some use spray headers that flush the plates with water. Of course the unit stays down (inoperable) until it is dried. In certain applications where dust is to be reclaimed, rapping hammers are used. This device is operated with compressed air to drive the rapping hammers. Under the filter unit is a hopper that opens into a reclaim barrel. You can see a system like this being used in a place where gold or silver dust was being trapped in the filter. Without a filter such as the electronic one, much of the precious metal dust would be lost.

Scrubbers

The scrubber can also be described as something similar to a water curtain. This system is widely used in laboratories for cleaning air that carries acid fumes, ammonia vapor, air-borne poisons, and foul odors. There is a deep rectangular box with a bottom drain. Inside this box are different plastic media filters, one behind the other. Water is sprayed on the front of the filters and spilled over the top. The different configurations of the media will impinge the dust, and the water will act as a barrier as well as rejuvenator and cleaner. The scrubber-filter is used in applications where the air is so saturated with pollutants that an activated charcoal filter could not keep up with the filter load. The down time of the air conditioning system due to filter service would be more than the actual running time.

In air conditioning, there are single-stage, two-stage, and three-stage filtration systems. In the single-stage system either a flat, disposable, pleated, or steel-mesh filter is used in the return air of the fan section. In the two-stage system, the flat, disposable, or pleated, filters are used as a pre-filter. The purpose of the pre-filter is to trap the larger and medium particulate matter. In doing this, you extend the life of the expensive final filters that are in the rack behind the pre-filters.

The final filters can be carbon, a bag, or pleated filter that has the extra depth to boost the efficiency to 55 percent, 85 percent, 95 percent depending upon the medium. In the three-stage filtration system, the pre-filter section will contain the flat, disposable, or pleated filters.

The intermediate filters will be carbon, charcoal, electronic, or extra-deep pleated in the 55, 85, 95 percent range. The intermediate filters will protect the very expensive after filters from loading with dust and extend their life for a very long time. The after filter will either be an absolute or magnamedia, depending on the manufacturer, with 99.9 percent efficiency.

Three-stage filtration will handle "clean rooms." If you have problems with pollen, allergies, asthma, or smog, try to get a job in a "clean room." Not only is the air clean, the humidity is controlled in the 32 to 36 percent relative with a temperature of 71 to 72 degrees F. The room has a positive pressure of about one-tenth inch water column, thus preventing contaminants from entering even when the doors are opened.

The slight positive pressure in the conditioned room keeps dust from migrating into the room through cracks around the doors, ceilings, or tiles. This is the ultimate condition when air conditioning a "clean room."

In the two-stage system, the manometer will normally read about two inches of water column. The three-stage system will read about three inches of water column.

Manometer

The manometer is a test instrument that has been mentioned several times. Some may not be familiar with what this device is, and how it works. The pressure in a sheet metal duct such as those used in ordinary air conditioning and ventilating systems is low. Air pressure of about 0.3 psi and less are common in these systems. Pressure gauges can be manufactured to read these pressures, but they are very costly. For this reason, a column of water is utilized. A pressure of 1.0 psi will support a column of water 27.7 inches high. This for example illustrates that a pressure of 0.05 psi would support a column of water 1.39 inches. In Fig. 22-6 a typical U-tube manometer is shown.

The inclined manometer is the one commonly found in air conditioning applications. This type is shown in Fig. 22-7 it is more sensitive to pressure changes. Pressures used in some duct work

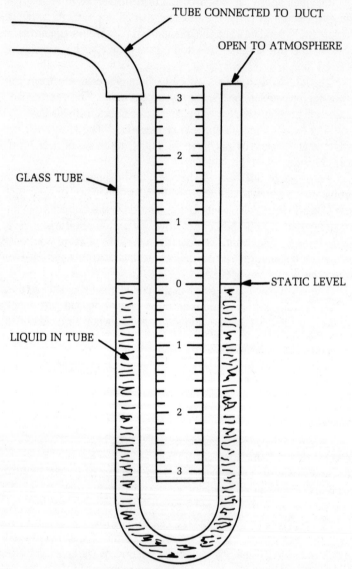

TUBE CONNECTED TO DUCT

OPEN TO ATMOSPHERE

GLASS TUBE

STATIC LEVEL

LIQUID IN TUBE

Fig. 22-6. U-tube manometer.

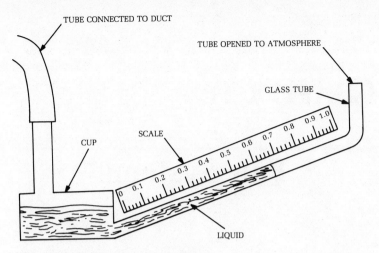

Fig. 22-7. Inclined manometer.

may be as low as one-tenth of an inch of water. The tube of this type of manometer is inclined at an angle that causes a slope of 10 to one; that is, if the tube is 10 inches long, its vertical rise would be one inch. If a pressure of 0.5 inches is applied to this manometer, the water would move a distance of five inches. Note the graduations on the inclined manometer, they are calibrated in psi.

CHAPTER 23

Fan Systems

In air conditioning, refrigeration, heating, and ventilation two types of fans are used, axial fans (propeller type) and centrifugal fans (squirrel cage). The axial fans usually have three or more blades. The blades are pitched at an angle to push certain amounts of air. The pitch ranges from 20 to 45 degrees. The centrifugal fans are divided into three categories; air foil backward curve, flat blade backward curve, and forward curve.

Axial fans have a shroud that fits around the blade and permits about a half-inch clearance all around. This shroud eliminates cavitation at the edges of the fan blade. A slapping effect would take place at the blade ends if clearance was too great. The axial fan blade is sized by its diameter, pitch of the blades (paddles), direction of rotation, hub bore size, and the number of blades.

The squirrel cage fan is enclosed in a housing called a scroll that fits around the fan. Its purpose is to direct and concentrate air flow. Squirrel cage fans are sized by diameter, width, cubic feet per minute of air flow (cfm), and the configuration of the blades. The air foil blade in Fig. 23-1 looks like the wing of a toy airplane. It will deliver five to seven percent more cfm per horsepower over the same backward curve flat blade. When the backward curve blade turns, the blades are swatting the air. Figure 23-2 shows a single width, single inlet, backward curve, flat blade, squirrel cage wheel. The arrows show the proper direction of rotation for the wheel. Figure 23-3 shows a double width, double inlet, backward curve, flat

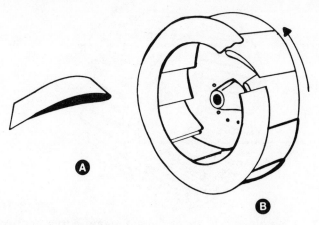

Fig. 23-1. Air-foil shaped fan blade.

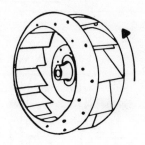

Fig. 23-2. Single-width, single-inlet, backward-curve, flat blade squirrel cage fan.

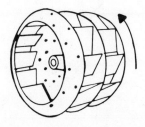

Fig. 23-3. Double-width, double-inlet, backward-curve, flat-blade squirrel-cage fan.

blade, squirrel cage wheel. The abbreviation for single width, single inlet wheel is swsi. The abbreviation for double width, double inlet wheel is dwdi.

On the dwdi wheel, the fan manufacturer will use two swsi wheels and mount them back to back with nuts and bolts, rivets or spot welding. Other manufacturers make a one piece dwdi wheel. Be very careful about the rotation when installing a new fan wheel. If rotation is wrong, won't function properly. You can recognize the

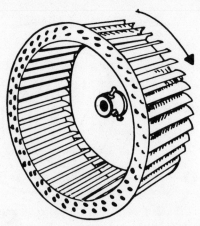

Fig. 23-4. Single-width, single-inlet, forward-curve, squirrel-cage fan.

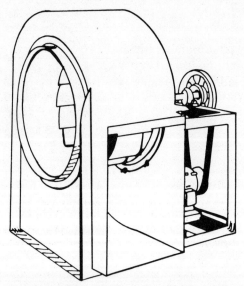

Fig. 23-5. Single-width, single-inlet, backward-curve fan.

proper rotation of a wheel either through the arrows or the type of wheel. This is also important when installing a new drive motor. If the rotation of the fan motor is wrong, the fan will not operate properly. Figure 23-4 is a single-width, single inlet, forward curve, squirrel cage fan. This fan is also made in double width and double inlet. Each individual blade is cupped like the palm of your hand. This throws the air in that direction. Figure 23-5 is a single width, single inlet fan. There is one wheel, and it is a backward curve. The round opening is the air inlet, low pressure side of the fan, and the rectangular opening is the discharge, high pressure side of the fan. These fans are weatherproof and may be used outside a building, inside a building, in attics or basements. This type of fan is also used for exhausting; however, if the fan scroll was insulated it could be used for air conditioning.

Figure 23-6 is a fan with double width, double inlet. This fan has two wheels that are backward curve, flat blades. The air inlet is on both sides of the shroud. The drive motor is located on one side with a pillow block bearing. This fan has two inlets and a single rectangular opening where the double wheel can be seen. This is a common discharge. The fan is a direct drive with a coupling. The motor shaft and fan shaft must be in perfect alignment or there will be undue stress on the bearings of each that will cause premature failure. The drive motor shaft and fan shaft are joined with a flanged coupling. The fan is mounted on rails with springs serving as vibration eliminators. The fan motor is drip-proof. The motor has air vent openings on the bottom of both end plates. The motor shown has a pedestal mount and permanent sealed bearings.

Figure 23-7 shows two views of the same fan. This is a double width, double inlet with wheels mounted back-to-back. Note the pillow block bearings, they have grease fittings on them. Fans such as these are usually continual run fans. They might be in operation 24 hours a day. For this reason the bearings should be lubricated every three months. These fans are industrial rated, heavy duty. Screens on both inlets are installed to stop any airborne debris. Pa-

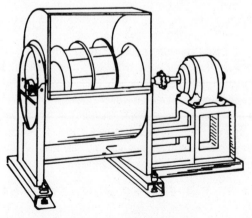

Fig. 23-6. Double-width, double-inlet fan.

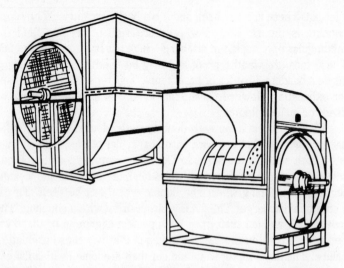

Fig. 23-7. Double-width, double-inlet fans, mounted back-to-back.

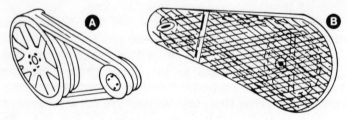

Fig. 23-8. Double-groove pulley arrangement.

per, styrofoam coffee cups, or anything else picked up or thrown into the fan will cause havoc with the balance.

Figure 23-8 is a typically pulley arrangement for a belt driven fan. Larger fans use double or triple grooved pulleys. Large fan blades are heavy and need more belt surface to move them from inertia without squealing. Belt guards should always be replaced for safety reasons. A belt thrown while traveling at a high speed is like a bull-whip and can do a lot of harm to a body. It is also there for your own protection if you should become careless. With the guard on you can't get caught in a moving belt.

RULES FOR PROPELLER FAN ROTATION

When the propeller fan blade turns, it exerts force through the fan shaft and applies that force directly against a thrust bearing.

Thrust bearings can be located at the ends of a motor shaft or pulley shaft. When the propeller fan blade turns in the proper direction of rotation, a large volume of air is moved. If the rotation is reversed, that air volume will diminish and the fan blade will churn the air instead of moving it.

Figure 23-9 is a propeller (axial) fan and shroud. The square, sheet metal enclosure around the fan blade is called a shroud. Note the arrows that show rotation of the blade. This particular fan cups the air and throws it forward, it is a heavy duty, industrial rated fan. The shroud and related metal parts are constructed from heavy gauge metal with steel structural reinforcement. Blades are supported by steel brackets. The fan blade is mounted on a driven pulley with the driver pulley being an integral component part of the assembly. The pulley of this fan adds rigidity and strength to the blade.

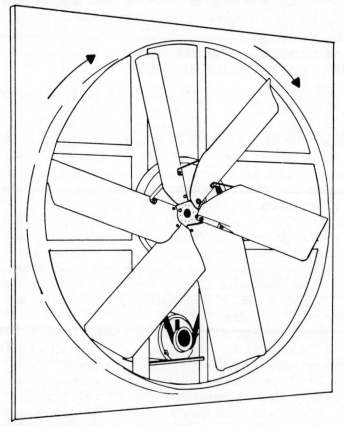

Fig. 23-9. Propeller (axial) fan blade and shroud.

A propeller fan blade that is equal in diameter to the forward curve squirrel cage fan delivers fewer cfm of air for the same speed. It doesn't make any difference how many paddles are on the blade or the degree of pitch. With the squirrel cage fan, air movement must be in the direction of the discharge opening in the scroll. In some larger commercial air handlers, multiple fans and scrolls are located on a common shaft. The scrolls will house dwdi cages. Some of the shafts can be made in two pieces and held together by a coupling. Other shafts are one piece. Both types of shafts are supported by either pillow block, flanged, or rigid mount bearings. The fan shaft on large units can be hollow or solid. Diameter of the shaft might reduce at its ends to accommodate stock pillow block bearings.

The maximum tip speed of a propeller or squirrel cage fan is reached when the fan causes the maximum amount of air movement. Further increase in the speed causes a decrease in air movement efficiency and only causes the fan to churn the air.

INCREASING CFM BY INCREASING MOTOR SPEED

Never place a load heavy enough to cause an electric motor to overdraw the amperage rating stated on its data plate. Whenever adjusting fan speed, use the clamp-on ammeter when fan is in operation to check current. Keep fan speed below tip speed. This condition can be observed by lower amounts of air supply accompanied by a decrease in the motor current. It is a good idea to adjust pulleys with a set amount of turns, checking amperage after each adjustment. Half-turns work well in most cases.

Listen for excessive noises and heavy vibrations. When exceeding the design speed of fan blades, bearings, and shafts, a vibration and harmonic sound can be created. Never overtighten a drive belt; it causes overload on the bearings and makes them operate at a higher temperature than they were designed for. This could cause grease to melt and drip from the bearing allowing it to operate with lack of lubrication.

The speed range of the forward curve wheels are one to 1500 rpm. The backward curve wheel operates in the range of one to 2800 rpm and might go as high as 3200 rpm in certain applications.

Electric current being supplied to forward curve fan drive motors will rise quickly with small increases in rpm of fan. The forward curve fan with the same wheel diameter as the backward curve fan will move more air at the same speed. Backward curve fan blades must turn much faster to move the same air volume as the forward curve fan. The advantage of backward curve wheels over forward curve wheels is that operating speed can be attained without a high rise in motor current.

When an additional heat load is added to an air-conditioning system, or more exhausting of air is needed, it might be necessary to increase the fan speed in order to increase the volume. When increasing speed, remember to check the motors operating current regularly. Make adjustments to pulleys in small increments such as a half turn at a time. Always make sure before adjusting the pulleys that any dampers in the system are open.

THROTTLING FANS

Air delivery from a fan can be throttled with dampers placed on the suction side of the fan. Do not use a damper on the discharge side of the fan, this can cause pressure build-up in the ducts and cause them to rupture. This is a common occurrence in fiberglass ductwork. When throttling the volume to slow, the motor current of the fan will drop. Less motor work means less motor current. Room registers in many cases have throttling dampers. Remember, these are on the discharge side of the fan. These dampers are to be used for air balancing. There are times when the owners start closing these off, not realizing the possibility of duct damage. In the average installation, air entering the room should be the same amount as the air returning to the unit. There are exceptions to this as was discussed earlier in the "clean room."

THROTTLING WATER PUMPS

As a point of information, air and water have been used to give examples of certain situations occurring in our industry. You might be involved with a water pump either on a tower or perhaps a lawn sprinkler system. Centrifugal water pumps are throttled on the discharge side only. That is the direct opposite of how fans are throttled. If a water pump has its suction side throttled, the pump might cavitate, lose prime, and/or burn out its seals. This will reduce motor amperage along with reduced flow of water. When the proper adjustment is reached by the technician, he should wire the valve handle to the position achieved so anyone else who follows won't open it full. In some installations where a water strainer or filter is being used, a pump can throttle itself, when this device becomes clogged. We relate back to the fan blade, this can occur in the same way to air movement if the filter in the system becomes restricted.

FAN VIBRATION

Check to see if there is a dirt build-up on the fan blade which can upset the balance of the wheel. Make sure balance weights haven't fallen from the fan wheel. Most fans are balanced at the factory. I can't impress upon you the importance of doing a thorough job if you decide to clean the fan cage. If you do not clean it entirely, you might end up with a worse vibration than you started with. This will be explained in detail in Chapter 30 on preventive maintenance.

With belt-driven fans, check the belt for cracks, pieces missing, and proper tension. A loose drive belt can set up a vibration like a string instrument, violin, or guitar.

Make sure all screws, nuts, and bolts are tight. Check the filter rack to make sure filters aren't rattling. If aluminum frames are used, a rattle can occur when the frame moves into a different position.

DIRECTION OF
DISCHARGE AIR AND FAN ROTATION

It is quite common to see, and I'm sure it has happened to many of you, a fan being taken apart to replace a motor. With the job complete, a service call comes in a couple days later complaining the unit is not working. A different man responds to the service call. He pulls his hair out finding that the man before him has installed the fan cage backwards causing the wrong rotation. With a direct drive fan, the task of taking everything apart begins. With belt drive fan motors, the work might not be as difficult. In Fig. 23-10 different fan variations are shown. Figure 23-11 shows the four most common drive arrangements for belt driven fans. In Fig. 23-12 a typical motor positioning is shown. You can see the rotation of the fan can be reversed by changing the mounting method. The shaft end of a motor is always considered the front of the motor. Facing the front of the motor you can see the direction of rotation. If the shaft rotates in the same direction as the hands on a clock, the motor is said to have a clockwise rotation. This is abbreviated cw (clockwise). If the rotation is in the opposite direction, it is said to have a counter clockwise rotation. This is abbreviated ccw (counter clockwise). Rotation will be designated with an arrow or one of the above abbreviations.

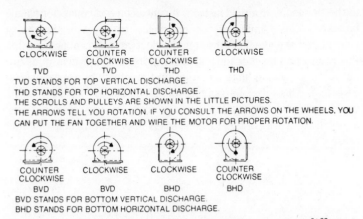

CLOCKWISE TVD   COUNTER CLOCKWISE TVD   COUNTER CLOCKWISE THD   CLOCKWISE THD

TVD STANDS FOR TOP VERTICAL DISCHARGE.
THD STANDS FOR TOP HORIZONTAL DISCHARGE.
THE SCROLLS AND PULLEYS ARE SHOWN IN THE LITTLE PICTURES.
THE ARROWS TELL YOU ROTATION. IF YOU CONSULT THE ARROWS ON THE WHEELS, YOU
CAN PUT THE FAN TOGETHER AND WIRE THE MOTOR FOR PROPER ROTATION.

COUNTER CLOCKWISE BVD   CLOCKWISE BVD   CLOCKWISE BHD   COUNTER CLOCKWISE BHD

BVD STANDS FOR BOTTOM VERTICAL DISCHARGE.
BHD STANDS FOR BOTTOM HORIZONTAL DISCHARGE.

Fig. 23-10. Mounting fan scrolls and rotation causing different directions in the supply air discharge.

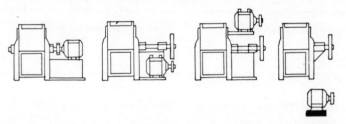

Fig. 23-11. Four most common drive arrangements for fans.

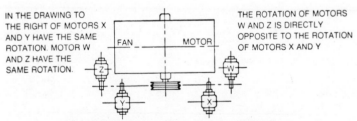

IN THE DRAWING TO THE RIGHT OF MOTORS X AND Y HAVE THE SAME ROTATION. MOTOR W AND Z HAVE THE SAME ROTATION.

FAN   MOTOR

THE ROTATION OF MOTORS W AND Z IS DIRECTLY OPPOSITE TO THE ROTATION OF MOTORS X AND Y

Fig. 23-12. Typical motor mounting positioning.

When a fan section is installed, it should be insulated against vibration. This is done in many ways. The cabinet can be mounted on springs, cork pads, or other types of vibration eliminator. This helps to stop any vibration being transmitted from the unit to the building structure. That is very effective in most cases. The next thing that has to be separated is the ductwork. Metal ductwork presents special problems in this area. Ductwork should not be joined

211

to the air handler in a rigid fashion. A canvas type of connector should be used. This eliminates the possibility of any vibration being transmitted and amplified through the ductwork. It is important that this connection be checked and not forgotten. The canvas does dry out and rot causing huge air leaks on the supply or return side of a system. This type of arrangement can be used on roof top units, ventilation fans, exhaust fans, package units, and any other that uses a metal ductwork. In most cases, glass fiber ducting helps to limit noise.

CHAPTER 24

Pulleys and Belts

All drive belts should be examined carefully when you are servicing a unit. A belt might look alright, but it can have a glaze on it or dry rot that causes cracking. Belt dressing can be used to prolong the life of a belt. Manufacturers use different numbering systems on their belts. There isn't a standardized numbering system.

Belts are measured by their width and circumference. Figure 24-1 shows how belt size is found when it is unknown. Hold a string on point one of pulley A, extend the string to point two of pulley B, and wrap string around pulley B to point three. Then continue from point three to point four and wrap the string around pulley A to point one. Then cut the string at that point. Take the string from the pulleys and measure the length with a ruler. If the string is 38 inches long, the perimeter is 38 inches. Measure the width of the pulley groove at its top, or measure the width of the worn belt at its widest point. Using this information, the proper belt can be installed. With the use of Fig. 24-2 and Fig. 24-3 you can see how belts are numbered for their size in both the automobile industry and the air conditioning industry, or I should say automotive belts and industrial type belts. Many times when a belt has been in service a long time, its numbers either fade or are worn off. The string method is an easy way to find the belt size.

Automotive air conditioning drive belts are built to certain specifications. Specify an air conditioning belt when replacing one in a motor vehicle. The belt has a steel wire in it, not fiber ones.

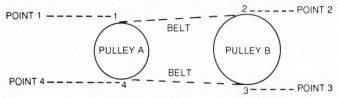

Fig. 24-1. How to measure a belt with string when the size is not known.

A green line or other coding color will distinguish belt as being used on automotive air conditioning systems. The conditions under which that belt operates are quite different than in a standard air conditioning system. Imagine the load on the belt when a car is traveling 55 miles per hour and the compressor clutch engages trying to create close to five tons of refrigeration. Of course the total capacity and load changes with the speed of the engine. The tonnage is rated in direct proportion to the engine speed. Due to the constant variance in speed, the special belt was developed. An ordinary belt would not operate under these conditions. The heat concentration under the hood of the engine compartment aids in the

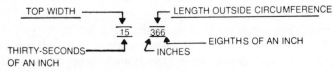

Fig. 24-2. Automotive belt sizing.

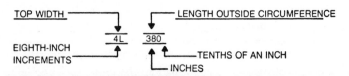

Fig. 24-3. Industrial belt sizing.

expansion factor of a belt, again the need for the steel ply.

A belt with size numbers 4L380, or 2380, or A38 might be confusing. In fact they are the same size. Look at Table 24-1 which is a width chart. It shows that 4L-2-A in each column indicates that each of the three belts has a width of ½-inch. The 38 or 380 in the size number reveals a circumference (perimeter) of 38 inches. The belts listed Table 24-1 chart three are heavy-duty industrial belts.

Table 24-1. Belt Width Chart.

WIDTH CHART #1	WIDTH CHART #2	WIDTH CHART #3
2L---¼ inch 3L---⅜ inch 4L---½ inch 5L---⅝ inch	0---¼ inch 1---⅜ inch 2---½ inch 3---⅝ inch	no # no # A---½ inch B---⅝ inch C---¾ inch

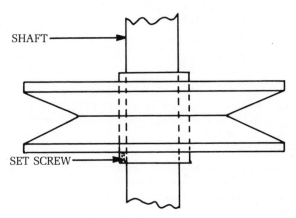

SHAFT

SET SCREW

Fig. 24-4. Fixed single pulley.

In Fig. 24-4 a single sheave pulley is shown. Note how it is sized, in case you must order one. In Fig. 24-5 you can see the difference of the fixed pulley as illustrated in Fig. 24-4 and the adjustable one shown. On small equipment, it is common that a single drive belt is used. When the equipment increases in size and horsepower, two or three and possibly more belts are used to drive large fans or compressors. Many older installations today are still operating with belt-driven compressors. In the previous chapter you learned that speeds can be increased or decreased when pulley size is changed. The capacity of the older compressors could be changed by pulley increase or reduction. With the newer equipment this is not possible due to the fixed-speed motor.

When you are working near drive belts, high speed pulleys, or any other moving machinery, never wear loose fitting clothing. Long sleeves, ties, or long hair can get pulled into the machine and cause you bodily injury.

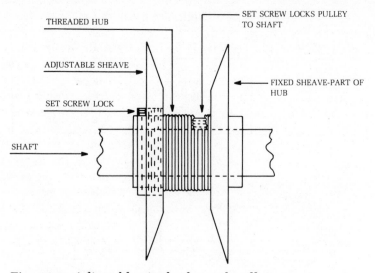

THREADED HUB

SET SCREW LOCKS PULLEY
TO SHAFT

ADJUSTABLE SHEAVE

FIXED SHEAVE-PART OF
HUB

SET SCREW LOCK

SHAFT

Fig. 24-5. Adjustable single-sheaved pulley.

BELT REPLACEMENT

I've seen many technicians install drive belts and not realize they were creating hidden problems for the future. Whenever you install a drive belt, whether it is being placed on a 1½ or 25-horsepower motor, use the same procedure. Overtightened belts destroy bearings and motor windings. The most common method for changing a drive belt is quick and doesn't require any other tools except a sturdy screwdriver. The screwdriver is used to roll the belt out of the groove of the pulley. The reason this method is usually detrimental to a motor is that the old belts have been in service for a long period of time. During the time the old belts were in use, it is safe to assume that the belts have been adjusted from time to time due to their stretch. Old vee belts also wear on the sides of the belt and makes them ride lower in the pulley groove which causes a loosening effect. Again the belt must be tightened. You can easily see that a 38-inch belt that has been in service for two years might have been adjusted a couple of times. This means the distance between the driver pulley on the fan motor and the driven pulley on the fan shaft has been increased. This distance might be one inch further than it was when the original belt was installed. When the original 38-inch belt is rolled out of the grooves in the pulley and the new 38-inch belt is rolled onto the pulley, the belt tension will be too tight. If a technician walks away from this installation leaving the adjustment as is, you have a pretty sure bet that eventually something is going to fail, probably the motor or fan. I've seen some technicians try to increase the speed of a motor-driven fan by tightening the belt. If you know the service history of a unit, as you would if you were a maintenance technician, then perhaps you can use the roll method to replace belts. In a situation such as that, belt replacement would be made instead of adjusting. You would know if the belt was tightened or not. There should be a little slack in the belt, between ¼ and ½ inch when pushed up and down.

The length of a belt will not affect the speed of the driven pulley; however, if the pulley is adjusted to make a drive belt fit, the speed will be affected. Many times another service problem is created by a service technician adjusting a pulley or changing one. It can cause icing of the evaporator and complaints that the unit it not cooling; to name just two situations created by an improper belt change.

FAN SPEEDS

Fans are designed to move a specific amount of air at a specific speed. This varies from one manufacturer to the next. Each has specifications of cfm of air over the coils. The motor manufacturer is the vendor to many air conditioning and refrigeration manufacturers. His motor has limitations as to maximum and minimum conditions for its operation. These limitations are listed on the plate that is attached to the motor. Two conditions that are deadly to the motor are exceeding current rating and exceeding the speed rating. Both of these violations will cause heat buildup in the motor and its pre-mature destruction.

I've mentioned the driven and driver pulleys. Let me explain that to those that might not understand. When a technician refers to a pulley, he must stipulate which one. Of course he can say, "the big one," or "the small one," but what happens when both pulleys are the same size? This is the reason for the driver being the pulley on the motor, the one that supplies the motivating energy. The pulley on the fan shaft or compressor is called the driven pulley.

As an example of this text, assume you have an electric motor with a speed rating of 750 rpm, with a maximum current rating of 10 amps. If the driver pulley is four inches in diameter and the driven pulley is four inches in diameter, the fan will turn 750 rpm. The r.a. (running amps) of the motor is measured at six amps with a clamp-on ammeter. If the driven pulley is reduced to two inches in diameter, the fan speed will be doubled (1500 rpm). For each rotation of the driver pulley, the driven has to make two rotations, causing the fan speed or whatever is being driven to double. Conversely, if the driver pulley is reduced, the speed of the driven pulley will be reduced. If you are trying to increase the speed of the driven pulley, for whatever reason, pay strict attention to the current load of the motor using your clamp-on ammeter. Fans have maximum speed for efficient operation. The amperage will increase with an increase of speed, due to the extra work being placed on the motor. When the peak fan curve is reached, cavitation takes place. Instead of the fan pushing air, it is only slapping it now, thus the amperage of the motor begins to drop. The fan has become inefficient eliminating the load on the electric motor. For this reason pulley adjustments should be made in small increments, each adjustment should be one-half turn, noting the starting point. If you have made two-and-a-half turns, and the fan becomes inefficient, you go back to two turns. If you find that no improvement is found, or the amperage overload is reached, the pulley can be re-set into its original position. Doing it this way will save you a lot of time and get the unit operational faster.

Adjusting the diameter of a single-sheave, adjustable pulley is an easy task. When making the same adjustment to a multiple-sheaved pulley the procedure is more difficult. In Fig. 24-6 a multiple-sheaved pulley is shown indicating points where measurements are taken when adjustments are made. When sheaves are moved further apart from each other, the diameter of the pulley surface, where the drive belt tracks, is decreased. When placed closer together, the opposite happens, the pulley becomes larger. If you are adjusting a double-sheaved pulley, two drive belts will be used and both of them must do equal work. If one belt is driving more load than the other, it has the same effect as if there were only a single pulley on the unit. The belt driving the entire load would continually fail prematurely. The only way load is distributed equally between belts is to adjust the pulley sheaves properly.

SHEAVE ALIGNMENT

This procedure is a simple one if the pulleys are of the same manufacture and design. On some equipment, the factory places different thickness driver and driven pulleys. This causes drive belts to track on a diagonal which causes excessive and premature wear of the belts. Simple alignment is done with the use of two methods. Figure 24-7 is self explanatory in the use of a straight edge to align

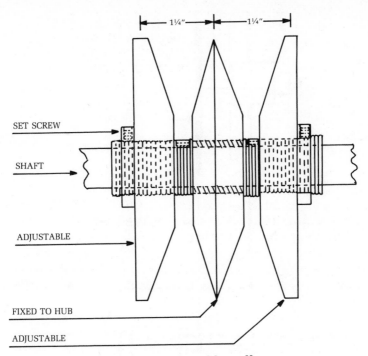

Fig. 24-6. Double-sheaved adjustable pulley.

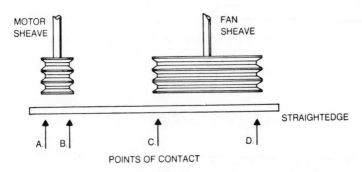

Fig. 24-7. Straight-edge method of pulley alignment.

pulleys. In Fig. 24-8 pulley alignment is shown being done with the use of a plumb bob. When a pulley needs replacement, three dimensions will have to be known. Diameter, bore, (shaft size) and width of groove.

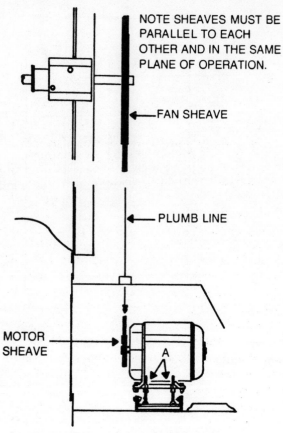

Fig. 24-8. Plumb-line method of pulley alignment.

CHAPTER 25

Leak Testing

One of the most frequent problems found in the field is that of a unit operating with an insufficient refrigerant charge or lack of a complete charge. In either case, the problem must be corrected. Leaks develop due to vibration or component failure caused by deterioration. Whatever the cause, there are no short-cuts in finding a leak and repairing it. Finding a leak requires more patience than anything else. The question always arises as to when and how long does a technician take to find a leak and repair it. This question must be put to the customer. If this is the initial call about low level of refrigerant, a very low volume leak might exist. Topping the charge once a year might be less expensive than trying to find the leak. Of course if the leak is right in front of you, fix it. I'm talking about a hidden leak. If the refrigerant leaks out quickly, the only alternative is to find it and repair it.

CONDENSING UNIT

It is usually the condensing unit where a leak test begins. The reason is that at this unit, a leak is the easiest to find and repair. The most important tool the technician has is his eyes. Look very carefully for signs of oil. Small amounts of oil attract dirt which make it very visible. Care should be exercised when handling used oil. In certain cases when a compressor has had a burn out, remember that acid is formed and carried by the oil. Heavy concentrations of acid can damage the skin. In addition to your regular hand tools, a few extra specialty items are needed for leak testing.

A bottle of nitrogen will be essential in some leak testing situations. Remember the nitrogen regulator; never use this gas without the proper regulator. Electronic leak detectors are good in isolating an area of the leak. In certain cases where the concentration of refrigerant fumes is very high, the electronic leak detector may be over sensitive to the condition and thus unable to pin-point the leak. A halide leak detector is a must. The concentration of the leak will be indicated by the color of the flame. Difficulties in leak testing with the halide might be encountered on windy days. The wind might carry away the refrigerant fumes faster than the halide leak detector can pick it up. For this reason I recommend you have a six-foot by six-foot piece of construction plastic to use as a tent. On windy days the plastic is used to cover the entire condensing unit. With the addition of refrigerant pressure to the unit, one corner of the plastic is lifted and the hose of the halide leak detector is placed under the plastic. The hose sniffs for the refrigerant. A concentration of refrigerant fumes should occur now that the wind cannot dissipate the leaking refrigerant.

One of the biggest mistakes a technician makes when using the halide leak detector is to use too large a flame in the chimney. It is only necessary for the tip of the flame to touch the underside of the detector disc.

At times, leaks develop and only show at high pressure. This is one of the reasons that nitrogen is used. On the data plate of most air conditioning condensing units, the test pressure is marked. This pressure is applied to the unit in the factory to make sure there aren't any leaks, defective joints, or materials. When applying nitrogen pressure, don't exceed these limits. The operating pressure of the unit is lower than those ratings on the data plate. Nitrogen should be added to the system up to the pressure at which the refrigerant of that system would operate. With the addition of the nitrogen pressure, the high pressure leaks can be detected without having the unit operate. Another thing to remember is the pressure of the unit on a really hot day would cause the units pressure to operate at a higher pressure. You might be looking for a leak on a cool day, and the pressure would not be high enough to detect the leak.

A product called Leak Detector is on the market. It is similar to the bubble solution a child makes bubbles with. If you do not have this solution available to you, ordinary liquid soap detergent used in the home will suffice. When the halide or electronic leak detector localize the refrigerant leak, the bubbles of the solution will form at the exact location of a leak.

EVAPORATOR UNIT

When a thorough testing of the condensing unit doesn't show a leak, move on to the evaporator section. Follow the same procedure as the one followed to test the condensing unit. You and the customer must be aware that leak testing of this nature is time consuming, and time is money. If the leak hasn't been found after the condensing unit and the evaporator section have been tested, it is time to talk to the owner again. Obviously the leak is somewhere in the piping that joins the condensing section and the evaporator section. This type of repair is going to take a judgment call on your part. If the piping is mostly concealed in the floors and walls, it might be best to run new piping rather than breaking down walls and tearing up floors. It would be advisable at this point to make an isolation test.

ISOLATION TEST

For the isolation test, you will need four line access Schraeder valves with pig-tails, a drum of nitrogen, and a nitrogen regulator. One feature about the inert gas nitrogen is the fact it is not affected by temperature changes as are refrigerants. The advantage of this in testing is that the pressure placed on the system will be the same when it is checked a day or two later, regardless of what the temperature is. In Fig. 25-1 the system is divided into four segments. Each segment is sealed with an access service valve hard soldered to it.

Nitrogran is then administered to each of the four segments with the use of your service manifold and gauges. Place the exact amount of pressure required in each segment. As stated before, do not exceed the specified pressure testing amounts that the manufacturer had placed on the unit. In an R-22 system, 350 psi pressure is sufficient; for R-12 system, 250 psi pressure is sufficient. The important thing to remember about this test is that the pressure amounts should be the same in the four segments. The nitrogen should be left in the system for several days, the longer the better.

When you return to the unit, place your manifold gauges on each segment and check the pressure. The unit with the leak will naturally have the least pressure. This isolation test is a costly one, but it definitely confirms which section of the system has the leak. Many times, this method reveals the leak to be in the condensing unit that

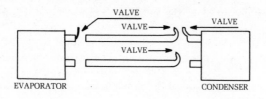

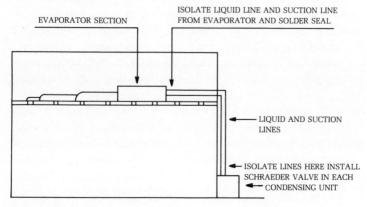

Fig. 25-1. Isolation leak test.

you checked in the beginning. This test eliminates tearing things apart for nothing. When separating the system for this test, remember you have to put it together again, so don't destroy piping where joints will be needed to place the system back together. If the leak is found to be in one of the lines connecting the condensing unit with the evaporator unit, a new line might be run instead of trying to repair the leak in the original one. This test gives you that option by eliminating any doubt as to where the leak is.

Tests like this one will detect a leak in an evaporator coil or condensing coil. You might then remove it from the unit and seal its ends so that it may be dipped in water, just as an inner tube is tested to find a leak. A leak in a coil might be located directly under one of the fins that is attached to the tubing. Water bubbles will show exactly where the leak was. Then the fins can be cut out of the way to make the repair.

You will be dealing with copper, steel, or aluminum when making your repair. You should be familiar with the three materials. In addition to silver solder and flux, it is advisable to carry a couple of brazing rods. A can of flux or pre-fluxed rods can be used if you store them in a dry place. Many steel parts are being used in the field such as condenser coils, receivers, accumulators, oil separators, to name a few. A rust hole or small crack can be repaired easily with the use of a brass brazing rod. The secret in any molten metal repair is to have the surfaces of the piece to be repaired clean and dry. Many coils are being epoxy coated at the factory to extend their service life. This must be sanded from the area to be repaired very thoroughly. Aluminum solder is used extensively in the field today. It is a special solder that needs very low heat. It is expensive, yet in comparison to the cost of a evaporator coil or condensing coil, the price of the solder is cheap. The use of this solder takes practice and is difficult to teach with a written word. Practice on an old coil first. Remember, the surface must be very clean, fluxed, and low heat must be moved constantly over the repair area. The use of the oxygen-acetylene torch is a must in this industry, and you must practice to be proficient at it.

Other means of repairing aluminum have appeared on the market, some work well. A compression type of fitting is being used where a joint of steel to aluminum is made. The device works fairly well. Another type of repair is coupling two pieces of aluminum with the use of a sleeve that is sealed in the line once in position by an epoxy. This adhesive is activated by placing heat on the fitting.

VACUUM PUMP

The vacuum pump is another important tool, without which professional results are impossible. If a system is only open to atmospheric pressure for a short period of time, a method called purging the system is used. This method uses the refrigerant itself to displace any air that might be in the system. The refrigerant being heavier than air displaces air. If a system has been open to the atmosphere for a long time, purging will not be satisfactory. It is in this situation that an efficient vacuum pump must be used. Purging is the introduction of refrigerant through the suction side of the system, letting it exit at the high side of the compressor. The refrigerant should be allowed to flow long enough for the air within to be driven out by the refrigerant.

The service manifold and gauges are hooked to the vacuum pump by the center hose on the manifold. The low-side hose hooks to the low-side of the system, and the high-side hose to the high-side of the system. The system should not have any pressure in it when the vacuum pump is being hooked to it. Any pressure in the system can cause the removal of oil from the vacuum pump. The low-side gauge should show pressure dropping a short time after the pump is placed into operation. A reading of approximately 28 to 29 inches of vacuum should be achieved in a short period of time. If vacuum only reaches 10 or 15 inches of vacuum, there is still a leak somewhere in the system. Both valves of the manifold are opened so the vacuum pump is drawing air from both sides of the system. If the vacuum doesn't drop, stop the pump operation. If there is a leak, the pump will draw ambient moisture-ladened air into the system. If the vacuum specified is reached, let the pump continue to operate for as long as possible, the longer the better. Overnight operation of the pump ensures a dry system. If this is not possible, maybe you can go to lunch and leave the pump on for a minimum of an hour. This will be alright for the average residential air conditioning system. On larger units, the 24-hour period might not be compromised when evacuating a system.

An important thing to remember about vacuum pumps is the need to change the oil frequently. The oil is a special light weight oil designed for vacuum pumps. If you find yourself in a position where you need pump oil due to the fact the pump fell on its side in the back of the truck and lost all its oil, type 300 refrigeration oil can be used. I suggest this only for an emergency and don't recommend it being used all the time in a vacuum pump. It will eventually induce pump failure.

By removing the air from a system, the moisture content is being removed also. I can't emphasize enough the importance of proper evacuation of a system before start-up. Not only does moisture cause icing within a system, it creates acid. Reducing the temperature to boil water is done by reducing the pressure. Water boils at 212 degrees F., at sea level where the pressure upon the water is 14.7 psia, (per square inch absolute). With the reduction of pressure upon the surface of the water or within the area, the boiling temperature is lowered. If a vacuum pump lowers the pressure to 0.12117 psia, (29.67 inches of mercury vacuum) water will boil at 40 degrees F.

Bearing those numbers in mind, remember that before you can effectively evacuate a low temperature freezer, the system temperature has to be above the 40 degree F. mark if you are going to use the conventional vacuum pump. Make sure there isn't any ice on the evaporator coil. If so, this will have to be defrosted. The defrost means can be anything from placing the unit into a defrost mode and letting its defrost heaters melt the ice or if the unit doesn't have a defrost mechanism, using a hair drier to melt the ice. A good solution if the product was removed from the freezer is to leave heat lamps inside the conditioned space to raise the temperature. The big "don't" is, don't use any sharp instruments to chip away the ice. Much damage has been done to many evaporators when people use a screwdriver or ice-pick as a defrost device.

The main things to remember about evacuation are that vacuum must be applied long enough to remove all the moisture from the system and to make sure the pressure is reduced enough so boiling of the water can take place.

With evacuation completed, close all valves on the service manifold. The center hose is removed from the exhaust port of the vacuum pump and then attached to a refrigerant drum. Refrigerant is allowed to flow into the hose from the drum. The manifold hose should be loosened at the manifold to purge all air out of the hose and make sure the hose is full of refrigerant. Open the low side hose and break the vacuum in the system with refrigerant. If the high-side gauge hose is hooked to a liquid service valve, the vacuum can be broken by allowing liquid to enter the system instead of vapor.

CHAPTER 26

Charging the System

Charging an air conditioning system or refrigeration system is a little more complicated than it looks. Many people buy a pound or two of refrigerant and place it into a system to get cool air out of it. Then they walk away from the unit only to have it fail a little while later. The service technician is then called only to give the owner the bad news of an expired compressor. In somewhat a state of shock, the owner says, "It was running a little while ago."

The proper amount of refrigerant and the proper type is important to the operation and life expectancy of a compressor. Units usually have the required refrigerant stamped on their data plates. The amounts can only be given if the unit is a package or a window unit. The capacity of split systems depends on the amount of piping between evaporator and condensing sections.

AIR CONDITIONING CHARGING TABLES

Many air conditioning manufacturers are attaching a charging chart inside their units. Table 26-1 is a typical chart. With the use of a few steps, this chart is easy to use and accomplish a correct refrigerant charge. The chart is used with both heat pumps and straight cool units. The unit must be in the right mode in order to use the chart properly. Table 26-1 is for the cooling mode. With thermometer, measure the ambient temperature and the suction pressure of the system during operation. Example; outdoor ambient temperature readings is 95 degrees F., dry bulb. The suction line

228

Table 26-1. Graphical Charging Chart.

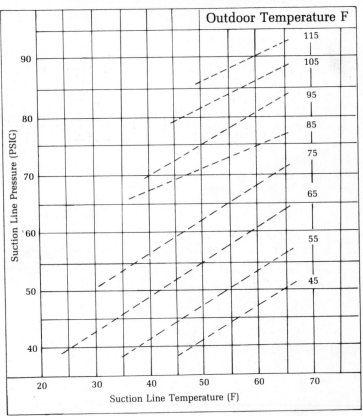

pressure is 70 psi. If you look to the right side of the chart, you will see temperature scales in degrees F. The lines that fall from the left of the numbers are the reference lines. Looking down diagonally to the left until the line intersects the 70-psi line. You can see that this intersection takes place on the 40-degree line. This should be the actual temperature of the suction line at the compressor. This temperature is measured with a thermometer. If suction temperature is higher than the reading, refrigerant should be added. If suction temperature is lower than the reading, some of the charge should be removed from the unit. In Table 26-1 a graph chart is used. Another manufacturer uses a charging chart as shown in Table 26-2.

The charging by feel method is fast becoming a thing of the past. It can be done by a technician with a lot of experience with a specific

Table 26-2. Charging Chart.

Outdoor Ambient °F	Suction Pressure at Compressor - PSIG														
	42	44	46	48	50	52	54	56	58	60	62	64	66	68	70
	Suction Line Temperature at Compressor °F														
Cool Mode Above 105													42	44	45
105												45	46	48	49
100											49	51	52	54	55
95											55	57	58	60	61
90										59	60	62	63	65	66
85									63	65	66	68	71	73	74
80									66	68	69	71	72	74	75
75							70	71	73	74	76	77	79	80	
70						71	73	74	76	77	79	80	82		
65						74	76	77	79	80	82	83			
Heat Mode 62					60	62	64	65	67	68	70				
57				55	56	58	60	61	63	64					
52			48	50	51	53	55	56	58						
47		39	41	42	44	45	47	49							
42	30	32	34	35	37	38	40								

Outdoor Ambient °F	Indoor Temp °F 60	70	80	Temp °F Change Across Indoor Coil
	Head PSIG at Hot Gas			
Heat Mode 37	191	222	254	25
32	184	212	244	23
27	175	202	233	21
22	169	192	222	19
17	158	182	211	17
12	150	174	201	15
7	144	166	194	13
2	135	158	183	11
– 3	129	150	175	9
– 8	122	144	168	7

manufacturer's equipment. Remember that most important of it all, the compressor should never have to operate continually with an amperage over its rated value. An ammeter should always be used when charging a system. Most air conditioning manufacturers design their units to operate with a 40-degrees F. evaporator.

SERVICE MANIFOLD AND GAUGES

The service manifold is fitted with two pressure gauges on top and at least two valves on its side or front. The average manifold has three hoses attached to it. These hoses should be kept in good condition, and when a hose shows signs of wear or leak it should be replaced. One gauge reads the low pressure side of the system.

On newer gauges, pressure scale is given in psi and metric kPa. Be sure you are reading the psi scale. On the pressure scale below zero is read as inches of vacuum. Most gauges, such as this one are usually incremented to 30 inches of vacuum. The outer scale reads pressure, the inner scales read temperatures. The three inner scales are temperature reading scales for, R-12, R-22, R-502 refrigerants. With the gauge constructed this way, you have the pressure-temperature chart in front of you as you charge a system. The other gauge reads the high pressure side of the system.

The three ports located at the bottom of the manifold are fitted with ¼-inch flare fittings. Service hoses have ¼-inch female flare connections. The hoses are equipped on both ends with rubber gaskets that should be changed when deterioration is noticed. These gaskets are responsible for the sealing of the fittings when opening a refrigeration system. Worn seals will allow the entrance of air into the system. The cost of these seals is negligible and a set of them should be carried in your toolbox. In certain cases where correct procedures are critical, ¼-inch copper is used instead of the rubber hoses. With flared connections on the copper tubing, the possibility of leaky rubber hoses is removed. In the case of evacuating a large commercial walk-in freezer where the most minute amount of moisture would cause havoc, copper tubing should be used on your service manifold. The extra time for making the hook-up is a good investment for proper operation of the freezer system.

Calibrate your gauges regularly. It is a simple thing to do, yet so many technicians forget to check them and have false readings from the equipment they are working on. All you need is a drum of refrigerant. Place the refrigerant drum in an area for 24 hours to allow the temperature of the refrigerant to assume that of the surrounding ambient of the drum. With the use of the temperature-pressure chart, it is easy to find the proper temperature of the refrigerant drum. Place your gauge onto the drum. The pressure should correspond to the amount shown by the temperature. It might not seem very important, but it is. If you are repairing an ice machine or a low temperature freezer, the difference of a few pounds of pressure can mean a machine is working or not. For that reason, check the calibration of your gauges from time to time.

SIGHT GLASSES

The sight glass shows the amount of refrigerant charge, and many have moisture indicators that show at a glance if moisture is in the system. The bad feature about the sight glass is that an inexperienced technician can overcharge a system trying to clear the bubbles or a flash from the sight glass. Those technicians learned incorrectly to charge a system until the sight glass is clear. If an ammeter was used during a load test, excess current would be apparent. If a system has a restriction in its filter drier, bubbles or flashing can occur in the sight glass. If a heat load is placed on a system, and it is operating at a temperature higher than its design temperature, flashing in the sight glass will occur. Example: you are checking the condensing unit of a 40-foot by 60-foot by 15-foot walk-in cooler. The sight glass is flashing. Unbeknown to you, the local beer vendor is delivering his weekly order of beer to the store. With a case of beer holding the door open so that he may enter and leave with his handtruck uninhibited. The evaporator fans are pulling air from outside the conditioned space over them. The hot moist air is causing a heat load on the evaporator higher than it was designed for. You can see in the situation above that the unit will seem to be operating low on refrigerant; however, as soon as the door is closed and the unit operates as it was designed to do, the sight glass will clear. What would happen to the excessive amount of refrigerant in the system if it had been charged during the above example? The unit would be overcharged and eventually, damage and failure would occur. If an ammeter is used, and it reads maximum amperage for the unit, there is something else wrong with the unit besides a low refrigerant charge.

CHARGING CYLINDER

When servicing certain appliances such as domestic refrigerators, window units, package units, ice machines, and automotive air conditioning, there cannot be any guess work as to how much refrigerant should be in the unit. There is a specific amount of refrigerant used in this type of equipment, and the manufacturer places that amount on the data plate. The amount needed in these units is usually given in pounds and ounces. When using small quantities such as this, one of the ways that it could be handled is with a very fine tuned scale or the charging cylinder. The cylinder is a valuable tool that should be a part of your tool inventory if you work on this type of equipment.

The operation of the charging cylinder is extremely easy once you understand the procedure. A tube where you place the amount of refrigerant is located in the center of the cylinder. On the outer perimeter is a plastic, movable, outer cylinder. At the top of the inner tube a valve and pressure gauge are located; and at the bottom there is a valve. Place a small amount of the refrigerant to be used into the cylinder through the lower valve. This small amount of liquid places pressure on the cylinder. A charging cylinder provides refrigerant in liquid form.

Read the amount of pressure on the gauge, rotate the outer cylinder until the proper refrigerant and pressure line up with the index. This outer, plastic, cylinder has graduations marked on it. These graduations are in ounces. Open the lower valve and allow liquid refrigerant to enter the inner tube, carefully observe the increments as the liquid column fills the center tube. When the proper amount of ounces is reached, close the lower valve. The exact amount of refrigerant needed to charge the specific system is enclosed in the center tube of the charging cylinder. Liquid refrigerant is dispensed from a refrigerant drum with the use of a valve labeled "liquid," or by turning the drum upside down. Always allow a couple of ounces of refrigerant to purge your charging hose. Never break the vacuum of a good evacuation until the charging hose is purged of air and full of refrigerant.

SERVICE VALVES

Depending on the unit you are working on, service valves vary and will be different. Figure 26-1 shows some different service valves. Service valves are devices that allow access to the sealed refrigeration system. The most economical way this is done is by crimping. This procedure is done at the factory. A small length of copper tubing is attached to the suction side and high side of the system. When the proper amount of refrigerant charge has been placed into the system, the tubing is crimped and soldered closed. When this is completed, there isn't a way to get back into the system. In Fig. 26-2 a crimped access is shown.

The next type service valve is called a line-tap-valve. Figure 26-3 shows one of these. This valve is used on soft copper and doesn't require soldering. It is installed with the use of screws. One of the disadvantages of this type of valve is its potential to leak. A small rubber "O" ring seals the valve to the refrigerant line and is the weak link in the chain. If used on a discharge line, the valve has a limited life due to the temperature variations being applied to the rubber "O" ring.

The most common service valve used on small equipment is the

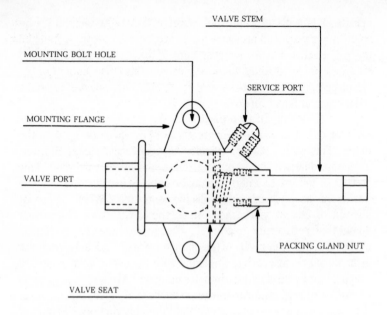

VALVE STEM

MOUNTING BOLT HOLE

SERVICE PORT

MOUNTING FLANGE

VALVE PORT

PACKING GLAND NUT

VALVE SEAT

NOTES
VALVE STEM BACKSEATED ALLOWS PASSAGE THROUGH VALVE
VALVE MOVED OFF BACKSEAT OPENS SERVICE PORT
VALVE FRONT SEATED SEALS PIPING AND ALLOWS FOR THE
    COMPRESSOR REMOVAL WITHOUT LOSS OF REFRIGERANT

Fig. 26-1. Service valves.

Schraeder valve. This valve is pictured in Fig. 26-4 and is very familiar to all for it is used on both the low and high sides of the
refrigeration system. Located in the hoses of your service manifolds
are indenters. These push the core of the Schraeder valve open when
the ¼-inch fitting is tightened onto the valve. The actual piece that
holds the pressure is the valve core. The valve cores are replaceable
with a valve extractor. Caution should be used when this type of
valve is used on a liquid line. Liquid refrigerant has a tendency to
spew when these valves are opened and closed. Burns will result
if refrigerant touches your skin. If a crimped type of seal is being
repaired, the Schraeder valve can be used. This valve can be acquired
with a small length of copper tubing soldered to it. These access
valves are soldered into place instead of crimped in the line.

The packed, angle-type valve is found in more expensive types
of refrigeration equipment and air-conditioning systems. As shown
in Fig. 26-5 it is entirely different from any of the previous valves

234

Fig. 26-2. Crimping.

mentioned. This valve has an actual valve stem that moves back and forth to open and close a valve from its seat. The valve stem passes through a gland nut on the outside of the valve. The packing within this gland nut keeps the refrigerant from leaking out of the system. A small wrench with a ratchet called a refrigeration service wrench

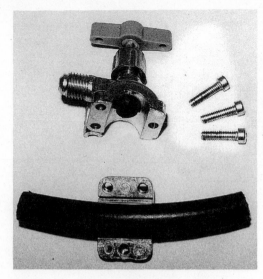

Fig. 26-3. Line tap valve.

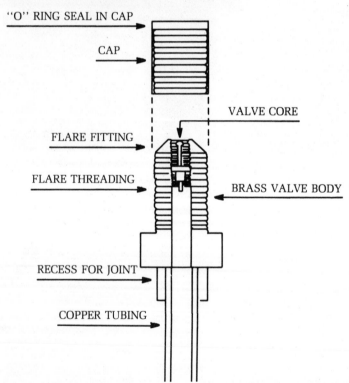

"O" RING SEAL IN CAP

CAP

VALVE CORE

FLARE FITTING

FLARE THREADING

BRASS VALVE BODY

RECESS FOR JOINT

COPPER TUBING

Fig. 26-4. Schraeder valve.

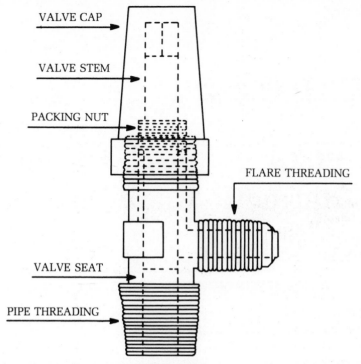

VALVE CAP

VALVE STEM

PACKING NUT

FLARE THREADING

VALVE SEAT

PIPE THREADING

Fig. 26-5. King valve, packing type.

is used to open and close this valve. On most semi-hermetic compressors this type of valve is used as the service valve and king valve on the receiver. Some versions of this valve require an Allen wrench, instead of a service wrench, for adjustment.

This type of valve is the safest to use since you have constant control of the refrigerant.

CHAPTER 27

Field Repairs of Semi-Hermetic Compressors

Semi-hermetic compressors are frequently used on light commercial air conditioning and refrigeration equipment. With the cast iron body and bolted head, oil pan, oil pump, and end cover that opens the electric motor section for service, certain field repairs can be made. Only two types of repairs should be attempted in the field. Replacement of valve plate assembly and/or the oil pump, if the unit has separate pump. No other type of repair should be attempted in the field.

The semi-hermetic compressor has many of the same component parts as an automobile. To rebuild a compressor thoroughly, special tools and test instruments are needed. If a compressor is not pumping properly, the head bolts and head should be removed carefully from the block. Looking into the cylinders, you can turn the crankshaft and see if all of the pistons move up and down. If there is a broken piston or connecting rod, put the head back on and replace the whole compressor. Some say that a compressor with the above condition would make a lot of noise. In some cases, this is so; however, I've come across many compressors that had broken rods or pistons and didn't make any noise.

If the pistons are alright, examine the valve seats and the valve reeds. Also, examine the head gaskets. There is one gasket between the block and the valve plate and another gasket between the valve

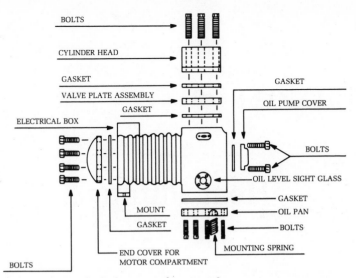

Fig. 27-1. Exploded view of a semi-hermetic compressor.

plate and the cylinder head. A blown head gasket can also cause an inefficient compressor.

Make sure that the old gasket is removed and all mating surfaces are clean. The cylinder head and the top of the block need the cleaning and take most of the time. Don't let the old scrapings of the gasket fall into the compressor body. Take time and be careful.

The valve plate assembly kit is usually designed to be used for many compressors. The only difference is among the gaskets used, and sometimes the reed valves are different. Use extreme care when matching the gaskets. They might look almost identical, but if you place one on top of another, you will see a slight difference in their cut. Make sure the reed valves and gaskets are exactly like the old ones that are taken out. Most of these semi-hermetic compressors have a service valve on both the suction and discharge sides of the compressor. Front seat both of the valves and the only refrigerant lost will be a small amount of vapor in the compressor.

When ordering a valve plate assembly, get all of the numbers that appear on the compressor, including model serial numbers. The vendor might want the model and serial number of the condensing unit also. Testing for an inefficient compressor is covered in the troubleshooting chapter. Figure 27-1 is an exploded view of a small semi-hermetic compressor.

CHAPTER 28

Heat Pumps

The heat pump is probably one of the most misunderstood pieces of equipment manufactured in the industry. I've heard customers and other technicians say that they weren't worth the time to make as far as heating goes. Of course this is the furthest thing from the truth. My first encounter with the heat pump theory was working with Frigidaire commercial air conditioning. The units were converted to heat pumps with the use of large solenoids and relays to coordinate the proper sequence of the solenoid valves that place the unit into the heating mode or cooling mode. They did in fact work very well.

Still not totally convinced, I attended a seminar conducted by Westinghouse Corporation that manufactured heat pump units for the government. A unit was produced for use in Alaska to prove the feasibility of the theory and its application. It too did work. With all this success with the heat pump, why can't you get Mrs. Jones' unit to work?

Before anyone can repair anything, he must know what it is supposed to do. In air conditioning theory, the unwanted heat in an objectionable space is transported and placed in an unobjectionable space. Before the new title of a heat pump, the unit was called a reverse-cycle, air conditioning unit. Think about that for a minute. What does the statement say? If a 4000 Btu unit is operating on the cooling cycle, the heat rejection from the condenser coil is being done outside the structure. By placing your hand in the discharge air of the condenser fan, you will feel the heat of compression combined with the heat from within the structure being transferred to the outside air. Imagine the effect if the unit was turned in the window placing the condenser section in the house. The heat of rejection is now being placed into the structure. That is the basis of reverse cycle, or the heat pump.

The skeptic among you is saying, "That's fine in the summer. But what about the winter." And that brings me to my next point and the Alaska Westinghouse seminar. When working with the refrigeration theory, technicians deal with fact, very definite facts that are scientifically provable. None of you can disagree with that statement.

For a moment, I want to go back to the very basics, the thing many forget, basics. What is heat? Everyone should know the answer to that. Heat is caused by friction from molecular motion. The slower the molecular motion, the less heat is created. With the advent of electromagnetic refrigeration, it has become a known fact that absolute zero, when there isn't any molecular motion, is found when the temperature reaches 460 degrees F below zero. Agreeing with the provable fact, why can't we heat a structure when the ambient outside the structure is 30, 10, or 10 below zero. The air is loaded with heat enough to do it, then why doesn't it work? If the theory is fact, and we know it is, then the problem has to be in the application. Westinghouse Corporation proved this beyond the shadow of a doubt. The problem then has to be in the equipment. Don't misunderstand that last statement, it is not my intention to demean any manufacturer, only deal with factual information. Look at Table 26-2, this is an example of a charging chart for a specific unit model. Look towards the bottom of the chart labeled heat mode. Notice how the discharge gas temperature drops as the outdoor ambient falls to −8 degrees F. The answer to the question "Why doesn't it work," then becomes a design factor to your locality. The theory is not at fault, the equipment is. Manufacturers produce their equipment for the larger air conditioning market. Their design figures are dedicated to this region only. You can see how expensive it would be to manufacture a special unit for each of the regional latitudes.

Now that you are aware of the theory and its workings, Fig. 28-1 is a typical heat pump. Engineering know-how eliminates the need to turn a unit around physically. The reversing of the action of a unit is done with certain component parts that are shown on the following pages. Figure 28-2 shows the unit with the component parts labeled for identification. Other units made by other

Fig. 28-1. Heat pump.

manufacturers will contain the same parts, but might be located differently. All heat pumps operate basically the same way.

The heart of the heat pump is still the compressor. The refrigerant routing must be changed; this is accomplished by the reversing valve. Figure 28-3 shows the reversing valve. This de-

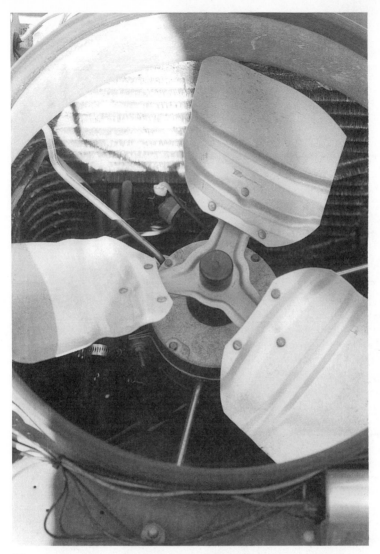

Fig. 28-2. Heat pump opened with component parts.

vice has made heat pumps manageable in comparison to the reversing action of the solenoid valves. In Fig. 28-4 the internal workings of the reversing valve are shown.

The solenoid pilot valve, that is mounted above the main valve body, actually causes the valve to shift. The pilot valve directs high

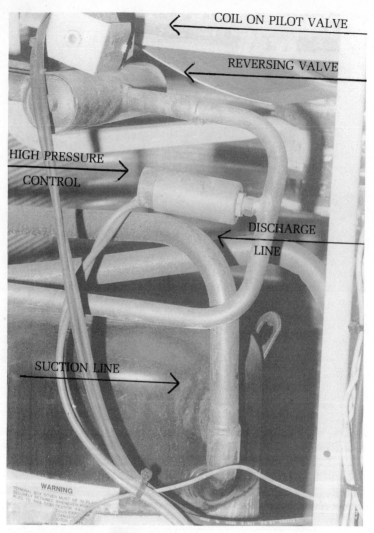

COIL ON PILOT VALVE

REVERSING VALVE

HIGH PRESSURE
CONTROL

DISCHARGE
LINE

SUCTION LINE

WARNING

Fig. 28-3. Reversing valve.

gas pressure to act upon the main valve spool, moving it laterally exposing valve ports allowing the refrigerant to change its path. The reversing valve is controlled electrically. Notice that the discharge inlet port of the reversing valve is by itself at the bottom of the valve. This port is always used for the inlet of the discharge gas. The valve routes the discharge gas. In the heating mode, the reversing valve routes the hot gas through the larger diameter pipe to the coil inside

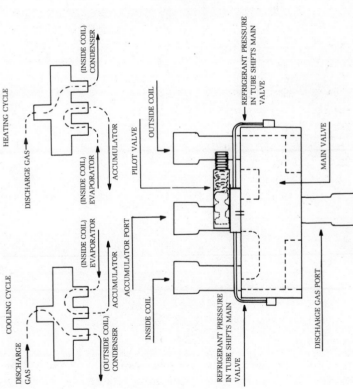

Fig. 28-4. Valve position for cooling cycle (upper left). Valve position for heating cycle (upper right). Note the differences in the routing of the refrigerant.

COOLING CYCLE

HEATING CYCLE

DISCHARGE GAS

DISCHARGE GAS

(INSIDE COIL) EVAPORATOR

(INSIDE COIL) CONDENSER

ACCUMULATOR

ACCUMULATOR

(OUTSIDE COIL) CONDENSER

(INSIDE COIL) EVAPORATOR

ACCUMULATOR PORT

PILOT VALVE

OUTSIDE COIL

REFRIGERANT PRESSURE IN TUBE SHIFTS MAIN VALVE

INSIDE COIL

MAIN VALVE

REFRIGERANT PRESSURE IN TUBE SHIFTS MAIN VALVE

DISCHARGE GAS PORT

the structure, the air handler. This section now becomes a condensing coil. Heat is rejected into the structure. With the cooling of the refrigerant, it goes into a liquid state and returns to the outside unit through the small pipe.

The liquid is fed through a metering device into the outside coil where it begins to boil and vaporize. The vapor is brought back to the compressor after passing through the reversing valve and an accumulator. The accumulator shown in Fig. 28-5 is in the system to avoid slugging the compressor with liquid refrigerant. The refrigerant is compressed and the discharge gas repeats the procedure. You can see how simple the heat pump really operates. Then what creates the problem? Ice, that is the problem. The outside coil becomes iced. This insulation of ice causes heat transfer to stop.

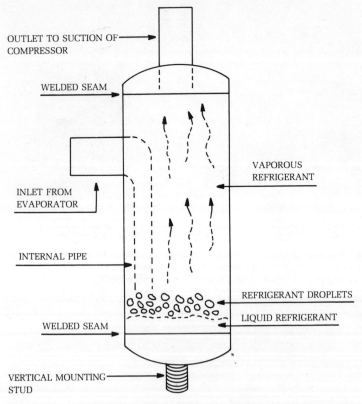

Fig. 28-5. Accumulator only allows vaporous refrigerant to return to the suction side of the compressor.

It is at this time the unit must defrost automatically just as most modern refrigerators do.

There are several controls used in the defrost cycle of a heat pump, and I think it is here that some get confused. A defrost thermostat is attached to the inlet pipe of the outside coil. When this thermostat senses ice, it closes and causes a timing device to actuate. Figure 28-6 is a defrost thermostat attached to the pipe.

The defrost timer can be an electronic circuit or an electrically-

Fig. 28-6. Defrost thermostat actuates a defrost cycle.

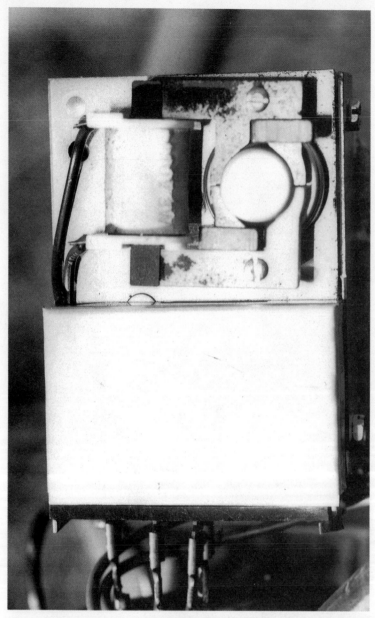

Fig. 28-7. Defrost timer holds unit in defrost cycle for programmed amount of time.

driven, mechanical device. They both do the same thing. They open and close switches. Don't forget the primary reason for the defrost, to melt the ice on the outside coil. This is done very easily by shifting the reversing valve, we allow discharge gas to melt the ice. Alright, that was easy except for one thing, the occupants are very uncomfortable with all of that cold air blowing out of the ductwork. The defrost timer shown in Fig. 28-7 has another job, and that is to energize the heat strips when the unit is in the defrost mode. The unit will stay in this mode until a signal is received from the defrost thermostat that the ice has been melted. At that time, the unit shifts back into the heating mode, reversing valve shifts again and the strip heat is turned off. What can be more simple than that?

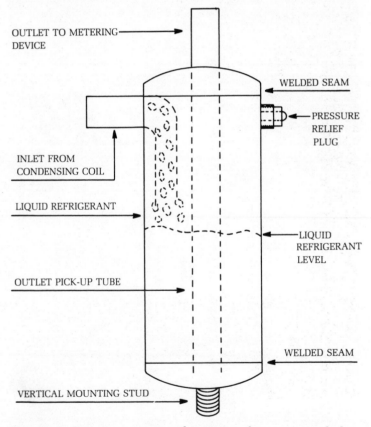

OUTLET TO METERING DEVICE

WELDED SEAM

PRESSURE RELIEF PLUG

INLET FROM CONDENSING COIL

LIQUID REFRIGERANT

LIQUID REFRIGERANT LEVEL

OUTLET PICK-UP TUBE

WELDED SEAM

VERTICAL MOUNTING STUD

Fig. 28-8. Receiver stores refrigerant when not needed to control load.

How often the unit goes into the defrost cycle depends upon the specific unit. You've learned in the basics that a coil will ice for one of two reasons. The unit is either low on refrigerant, or the coil has insufficient air passing over it. I think most of the time the second problem will exist. It could be a design flaw or an outside problem being introduced.

A receiver should be used on the heat pump because of the different amounts of refrigerant being used in the cooling mode and the heating mode. In Fig. 28-8 the inside of the receiver is shown. Although the receiver looks very similar to the accumulator, their functions are totally different as is their construction.

Figure 28-9 shows the refrigerant flow when the system is in the heating mode. In Fig. 28-10 the refrigerant flow is shown when the system is in the cooling mode.

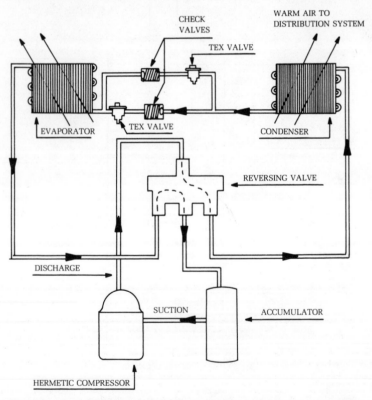

Fig. 28-9. Refrigerant flow through system in heating mode (black arrowheads indicate direction).

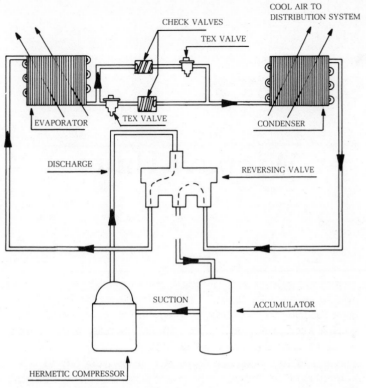

Fig. 28-10. Refrigerant flow through system in cooling mode (black arrowheads indicate direction).

CHAPTER 29

Window Units

Window units are the answer to those structures that do not have ample room for a duct system, or where the cost would be exorbitant to install a central air conditioning system. The reasons are as many as the manufacturers of window units. A variety of sizes have appeared on the market. Sizes range from approximately 4000 Btu up to 36000 Btu. Of course the larger units are mounted through a wall not a window, however, they are referred to by many as window units.

The window unit is a complete packaged air conditioning system contained within one cabinet. They can be purchased as straight cooling units, or heat pumps, or with electric heat. Window units are made to fit standard sash window frames, aluminum window frames, casement window frames, and to be mounted through walls. Installation instructions can be acquired from the distributor or the manufacturer.

Small window units have a case built around them. In the illustrations of this chapter, a small unit is being used. A unit such as this, when removed from the window, has the case (metal cabinet) attached. The larger units have a sleeve, or shell, that remains in the opening. The air conditioning unit components are built upon a heavy steel pan that is called by many in the industry, the chassis. Usually, the shell is attached to the building structure with screws, making it stable in its support of the weight of the air conditioner. There are rails at either side of the bottom of the shell for the chassis to slide on. When the chassis is in operating position in the shell, several screws are placed in positions that lock the chassis in the shell. This is done for security reasons. When locked in position, the chassis can't be slid into the house which might leave a large enough hole for a person to climb through.

Approaching the window unit from the front, there is always some type of attractive front cover. There may be a small door in it that conceals the control knobs. This panel must be removed. They are either fastened with knurl nuts located on the sides of the unit, or a spring-type fastening device. With the front cover removed, the filter can be seen. These are usually made from a thin sponge or synthetic material. It is placed across the face of the evaporator or fan intake, depending on the unit. An electrical service cord enters the chassis. At that position there is a metal plate held in place by screws. With the electrical service cord removed from the receptacle (wall outlet), remove the screws and metal plate. This should expose the control switch and sometimes the run capacitors.

There is nothing else that you can get to, without now pulling the unit out of the shell, or the entire assembly from the window or wall. With some units, you can pull the chassis out far enough on the shell's rails to get to the component you have to work on. If not, arrange a place to put the unit when it is removed from the wall. A homeowner might get a little upset if you put the unit in the middle of a highly polished wooden dining room table. Make sure there is a work area suitable for the task. Usually, a window unit larger than a 8000 Btu rating, should be handled by two people, not only because of the weight but also due to the awkward shape and weight distribution of the unit.

With the unit removed and placed on a stable work surface, you will see the compressor immediately. Figure 29-1 shows the case being opened. The components are exposed when the case is removed. In Fig. 29-2, the compressor is seen. It is a hermetically-sealed unit with either a reciprocal type or rotary type of compressor. The rotary compressors are known for their high-pitched noise. Notice the compressor terminal box and cover, Fig. 29-3.

Typical window units use a double-shaft, fan motor. It is usually mounted directly to the cabinet that housed the fan scroll, or on a pedestal, as seen in Fig. 29-4. The pedestal, if used, is either bolted or welded to the chassis. This double-shaft motor drives an evaporator, squirrel-cage fan on the inside of a fan scroll, and an axial condenser fan blade in a shroud. Many window units are designed with a ring surrounding the axial fan. This is called a slinger.

Fig. 29-1. Window unit with outer case removed.

Condensate water is picked up by this ring and slung onto the condensing coil to help cool the refrigerant gas. Some of you are asking about icing water in the case of a heat pump. Good question! Many units are designed with a water flap valve in the bottom of the chassis near the outside fan. It is thermostatically controlled so that the water valve opens with a low ambient temperature. This means that no water will collect in the pan on cold days because of the large drain hole. Also found on or near the fan motor will be the fan run capacitor. Remember that on a small unit with a small fan motor, there might not be a run capacitor if the motor is a split-phase or a shaded-pole type.

Some units have a manual control for a damper that opens to allow fresh air into the fan scroll. There is also a metal mesh to stop small objects from entering the squirrel cage fan scroll. Figure 29-5 shows the axial fan blade.

In Fig. 29-6, the scroll cover is being removed to expose the squirrel-cage fan blade. In Fig. 29-7, the squirrel-cage fan blade is in view. The squirrel-cage fan blades are fixed to the fan shaft with the use of one of two methods. Some units have the fan in front where you can reach into the fan and get to the hub. The hub either

Fig. 29-2. Hermetic compressor.

has a surface locking bolt or a recessed Allen set screw. Other units where the fan hub is not accessible, the blade directly opposite the set screw has a cut-out in it. A very long Allen wrench is used in this case to loosen the set screw. This process will be used if the fan drive motor has to be replaced, or the squirrel cage fan has to be removed for either cleaning or replacement.

Fig. 29-3. Terminal box with cover removed.

I've found the majority of service calls made on window units are,

■ Units have not been cleaned. Evaporator coils restricted to a point that no air can pass through it. Icing of the coil is very common. Without a low pressure control, this units compressor slugs causing possible damage.

■ Fan motor inoperative, causing the same effect as stated above, icing. The fan motor might have been saved if it had provisions for lubrication, but never received it. Bearing failure is the number one motor problem in window units. When the bearings fail, the rotor falls from center and causes heat generation or actual physical contact with the stator.

256

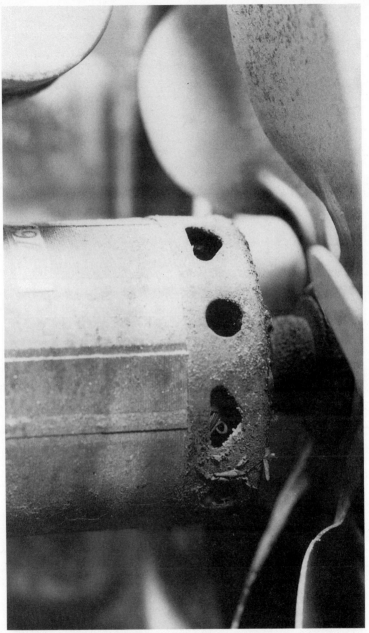

Fig. 29-4. Double-shaft fan motor drives both the condenser fan and evaporator fan.

Fig. 29-5. Axial condenser fan blade.

■ Wiring at compressor corrodes and separates from terminal.
■ Leaks
■ Compressor failure
■ Other—control switches, capacitors, etc.

Replacement of the fan drive motor in my opinion is the most difficult service task a technician can have on a window unit. Most would rather change the compressor. The reason is rusted nuts and bolts and tight places to work in. First get some penetrating oil. Spray

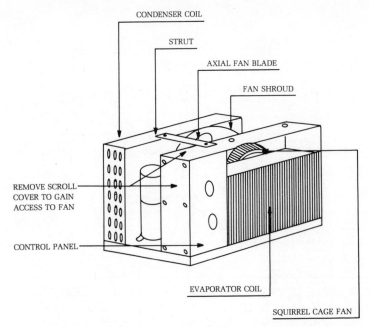

CONDENSER COIL

STRUT

AXIAL FAN BLADE

FAN SHROUD

REMOVE SCROLL
COVER TO GAIN
ACCESS TO FAN

CONTROL PANEL

EVAPORATOR COIL

SQUIRREL CAGE FAN

Fig. 29-6. Removing the fan scroll cover exposes the evaporator fan.

all the nuts and bolts and don't forget the fan set screws. Usually the problem area is going to be the removal of the axial condenser fan blade. This part of the motor shaft is exposed to both the weather and possibly condensate water that is slung by the fan slinger. This means there is a possible bonding between the fan hub and the motor shaft. With some blades, you can't manhandle them or apply heat to them due to the fact they have a rubber insert in the fan hub. At this point, I would check for the availability of a new axial fan blade.

In most cases of fan motor removal, the condensing coil is taken loose from its mounts. The refrigerant piping is not disconnected. The pipes, even if they look pitted, usually have enough give to them that they will flex without any problems. With the condensing coil moved to one side, the fan motor can be removed. The squirrel-cage fan blade stays in the scroll and can be removed after the motor. The blade should be cleaned at this time. In fact the whole unit should be washed with a good commercial detergent or some industrial coil cleaner. Of course be careful of the electrical components. Keep the water and cleaner from them. If this were a cleaning instead of a motor change, you should cover the motor with plastic to keep

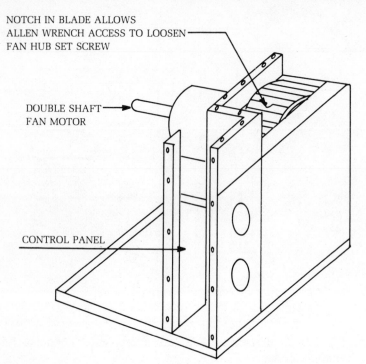

NOTCH IN BLADE ALLOWS
ALLEN WRENCH ACCESS TO LOOSEN
FAN HUB SET SCREW

DOUBLE SHAFT
FAN MOTOR

CONTROL PANEL

Fig. 29-7. Squirrel-cage fan blade.

the water from it. Install the new motor and possibly the fan blade in the reverse order. If a new fan blade is being installed, check for proper rotation. The same applies to the fan drive motor. Most of these motors are directional depending which shaft goes where. Be careful about rotation.

With the reassembly of the unit, I whole-heartedly recommend all of the nuts and bolts be coated with a good quality waterproof grease or compound of the same nature. This makes life easier, the next time you have to disassemble the unit. Technician and homeowner alike, make a note that even brand new parts have a record of quick failure. The motor might have a warranty period of 90 days or possibly a year. A few months of wet weather can cause deterioration of nuts and bolts by rust. Don't forget to protect the new fan motor shaft, remember the battle getting the blade off.

CHAPTER 30

Preventive Maintenance

Both large and small service companies and dealerships have discovered through their service contract program that a good preventive maintenance program is a money saver. Service contract customers pay their annual fee, then it is up to the service company to keep the unit operational at all time regardless of cost. This is why a comprehensive preventive maintenance program is important to you. With a program in place your conscientious efforts will keep whatever you are working on operational longer without costly breakdowns.

Most programs are performed on a semi-annual basis. With two inspections a year, the equipment should be alright. Have a set time in the spring to check out the air conditioning system and in the fall to check out the heating system. The procedure that follows is geared for the split system. Package units are serviced in the same way with some extra caution and variances. To begin with, service the condensing unit first.

CONDENSING UNIT

Walking towards the outside condensing unit, keep a watchful eye open for things like overgrown hedges or foliage. See if the unit has settled into the earth, if so, it must be raised. Of course those of you that are working on a condenser unit that is elevated will not have that problem. Next thing you do is open the power source disconnect. You don't want any electricity flowing into the unit from this point on. Clean the area around the condensing unit for about three feet in all directions. This includes anything that is growing or stored that close to the unit. Remove the service access panels and the top of the unit if it is designed that way. You should now see the fan motor and blade, compressor, condenser coil, and the other components. Cover the condenser fan motor with a plastic wrap to make sure water cannot enter the motor. Bring over the water hose with a nozzle. Don't direct the stream of water with too much pressure or it will bend the aluminum fins of the condenser coil. Start from the inside of the unit, directing the stream of water through the coil toward the outside of the unit. Dirt and even grass cuttings will be driven out of the coil by the water pressure. Then wash the coil in the reverse direction. This is the reason the motor should be covered. If the coil is really dirty, you now apply a detergent with the use of a spray bottle. The coil should be wet. Let the detergent stay on the coil for the time it would take to have a cup of coffee. Both sides of the coil should be soaked with detergent.

Flush the detergent from the coil with the water hose, again from the inside out. The water should actually be directed in the opposite way the air flows through the coil. Remember that the air discharge can be vertical, horizontal, and at a diagonal. Make sure that the coil is clean. A little extra time now makes the unit operate more efficiently.

With the condenser coil clean, take the wrap off the motor. Examine the motor carefully. If there are oilers on the motor place several drops of electric motor oil into each oiler. If you don't have electric motor oil, use #30 non-detergent motor oil. The oil holes might be plugged with plastic, lead, or a spring loaded cap. This has to be moved for the oil to be placed into the motor. Waterproof grease should be used for the fan shaft to keep it from rusting allowing it to be removed easily if need be. Check visually the receiver, accumulator, driers, and any other steel components for signs of rusting and deterioration. A little sandpaper and paint at this time will save time and effort later on. A good epoxy paint should be used.

COMPRESSOR

Open the compressor terminal box and check the compressor wiring for tightness. Make sure the connections are tight. A spray product should be used to displace moisture and coat the metal with a protective coating. Make sure that the product is approved for use on electrical circuits. It must be a non-conductive material.

After replacing the terminal cover on the compressor, open the control panel of the unit. Check all connections for tightness. Spray the panel components with a liberal coating. Replace the cover on the control panel, and replace all the service access panels.

Place your service manifold and gauges on the valve ports in order to take pressure readings. Turn the unit on in the cooling mode. Wait several minutes until the water that may be left on the coil dries. Take the pressure readings and record them on a piece of paper. You should have a good idea if they are within normal ranges for the ambient temperature of the day you are doing the cleaning.

EVAPORATOR

Turn the unit off. Now you have to service the air handler or evaporator section. This part of the system will require special attention depending upon where it is located. If an air handler is located in an attic above a finished ceiling of a living space, water damage from overflowing condensate pans during the cleaning process can cause extensive damage. The use of a water hose might not be advisable in this situation. A pressurized container such as those used by lawn maintenance people to fertilizer shrubbery is fine for this job. This type of sprayer is used with the detergent first, then flushed with clean water, it is used to rinse the evaporator coil. The amount of water can be controlled this way allowing time for the condensate drain to remove the waste water.

The first step with the evaporator section is to remove access covers. Cover the evaporator drive fan with plastic, if necessary. Clean the evaporator coil in the same manner you cleaned the condenser coil outside. The evaporator coil could become really dirty with compacted dirt lodged deeply in the fins of the coil. A drop light, held on one side of the coil with you looking through it from the other side, will show you immediately if the coil is dirty and where. The light will only shine through and be seen, where the air can pass through the coil. Wherever the light doesn't shine through will need extra cleaning due to the blockage. Many times when checking the operating pressures at the condensing unit, a low pressure reading is sometimes interpreted as a low refrigerant level. The unit should be checked first for filter or coil restriction before any refrigerant is added to the system. If overcharged, the system has more of a chance to damage the compressor by refrigerant slugging.

It is very common for units that have been in service for several years to have dirty squirrel-cage fans. The curvature of each individual blade can become filled with dust and dirt. It might be necessary to remove the fan assembly from the unit to facilitate cleaning. The hose might have to be used outside the structure. If you attempt to clean the fan in the unit, exercise extreme caution when both cleaning and starting the unit after cleaning. Any loose dust or dirt might be blown through the ductwork onto the customers expensive furniture. Place something on the discharge side of the ductwork to trap all dirt that might become airborne.

Depending upon how the drive motor is mounted, the fan drive motor should be oiled with the same oil as used on the condenser fan drive motor. Some motors are sealed and do not require lubrication. Check closely to make sure. If a drive belt is used, that should be examined carefully. Check the edges for glazing condition. Check for cracking or signs of dry rot deterioration or loose fabric wrap. Bending a belt in your hand will crack it if dry rot is in belt. If in doubt, replace it rather than get a service call on the weekend.

One of the most important things to check in the evaporator section is the condensate pan. Algae and rust always clog condensate pans and drains if allowed to. Manual removal of this material is recommended rather than flushing most of it down the drain line. If there is a large rust chip, it can become lodged at one of the fittings and cause future stoppage of the drain line. Algacide tablets do a fine job if placed in the pan properly after cleaning. This tablet retards the future growth. Flush the drain line with bleach. Some units that are installed in the attic or over a trap door in the garage might have a filter frame located in the return air duct. Some units have operated with this filter until it became so restricted a service call was required. The owner changed the one in the hall ceiling all the time but never knew of the one in the attic, so check it.

Much of the preventive maintenance can be done by the owner if he so chooses. If they are not agile enough then a service contract should be placed upon the unit, and it should be checked twice a year. When checking the heating cycle, examine the heat strip fuses and terminal tightness. The unit should be cycled into heating to check the reversing valve operation and the heat strip amperage draw to make sure it is within specification for the number of resistive heaters.

With the use of the droplight and flashlight, examine the entire length of the drain line in the attic if possible. Make sure that it is secure. When pvc piping is used, this is especially important for it has the tendency to soften and bend with the heat if not supported properly. In many cases the bending of the pvc causes water trapping in the pipe. If the condensate pan is shallow and there is not enough water weight to push through the trap, overflow occurs and a ceiling is lost. Never double trap a condensate line for this reason. In many installations where the pipe exits the attic and flows down a wall, the lowest point in the drain would be the best location for the trap.

THERMOSTAT

The last thing to service, and it only takes a few minutes to do, is the thermostat. It should be checked for level or plumb depending upon whether it is square or round. This was discussed in Chapter 11. Thermostat sub-bases have a tendency in certain installations to become loose and throw the whole calibration off. This happens from vibrations to the wall, or from a person constantly turning the switches or temperature settings.

The whole procedure should only take a couple of hours depending how dirty the system is. Of course this time can drop to an hour if things are located in accessible locations.

CHAPTER 31

Troubleshooting

We've all heard the expression, "parts changer." I've seen mechanics that could fix anything; however, it took a different person to tell the mechanic what to fix. I'll never forget the story of the man that disassembled an appliance completely. While standing there looking at it and scratching his head in bewilderment as to what was wrong, his eyes locked on to something that brought horror to his face. The plug had not been in the receptacle and could not supply electricity to the appliance. The moral of this story is, diagnose the problem when troubleshooting.

Every technician has his own way of doing things, yet as busy as his day goes with sweat pouring into his eyes, and the list of service calls growing, he will systematically diagnose a unit's problem. Stop, look, and listen, before you open your toolbox.

Although there are solutions to a problem, all experienced technicians agree that you must start from an end. In the troubleshooting procedure, you cannot start from the middle somewhere and work towards an end. This wastes time and can lead to false repairs.

I am presenting this chapter in a way that will be helpful not only to the technician but also to the novice owner of an air conditioning system. The problem will be stated, followed by possible solutions.

CONDENSING UNIT
AND EVAPORATOR NOT OPERATING

■ Check the settings on the thermostat. Is unit turned on and is the temperature high enough to be calling for the unit to operate.

■ If unit fails to come on, turn fan control switch of the sub-base to the constant run position. If the fan comes on, low voltage is present in the system. If it doesn't come on, there is no low voltage in the system. Check at the transformer for secondary voltage. If not present, check the primary.

No primary voltage—   Check fuse that supplies the unit.

No secondary voltage—   Bad transformer.

Both primary and secondary voltage present—
                           Check at sub-base of thermostat
                           for 24 volts.
Voltage present at sub-base—
                           Replace sub-base and thermostat.
Voltage not present at sub-base—
                           Control wiring from transformer to
                           thermostat defective.

EVAPORATOR SECTION BLOWING WARM AIR

■ Check condensing unit to see if it is operating. Place one hand on the liquid line and the other on suction line. If the condenser fan is operational and there isn't any difference in the temperature of the two lines, the compressor is not operating. If the fan motor is not operating, check for line voltage and low voltage in the control panel. You must establish if the problem is in line voltage or low voltage. If line voltage is present, and there is no low voltage, nothing will operate.

No line voltage—   Check fuses for the supply voltage
                           to the condensing unit.

No low voltage—   If present through the thermostat,
                           a wire from the thermostat to the
                           condensing unit is open.

Both voltages present—   The compressor contactor on most units acts as a switch to supply power to the fan and compressor. Check for bad holding coil or bad contacts.

Fan operational, no compressor—

Check terminal box of the compressor to make sure voltage is present at the compressor. Could be burnt wire or bad compressor. Check compressor. Hot compressor could have open overload relay.

It is at this point, that proper diagnosis be made. How stupid you will feel when you replace a part only to find out that the compressor is defective. You have to know at this time if the compressor has to be replaced or not. This is the most expensive part to replace. Follow the compressor tests mentioned in Chapter 3. Only after you are totally convinced that the compressor is not electrically defective, can you continue. Place your charging manifold and gauges on the service valves. If pressures are low with the unit off, there might be a refrigerant problem. This could be why a compressor overheated, lack of cool suction gas. Open service disconnect. With ohmmeter, test continuity of all the controls leading to the holding coil of the compressor contactor. It is not uncommon to find a control switch that has opened and failed to reset. The control became defective either from use or defective manufacture.

Remember to stop, look, and listen. If you find a unit without refrigerant, and the leak test shows an open pressure relief valve or ruptured line, along with an open high pressure control, something must be wrong. Don't just repair the leak, charge the unit, and run. The possibility of your being called back is great. Something such as the conditions I described can be caused by a couple of things. First, what is the cause of extreme high pressure? Something lodged in the discharge would cause high pressure. Intermittent operation of the condenser fan motor would cause the problem. You can see there are several possibilities for the probable cause. Explain this to the customer. You might have to place recording monitor equipment on the system to record pressures and amperages. This is a costly test due to the equipment used and the time consumed.

With the use of a pressure recorder and an amperage recorder, tests can be made over a period of time. In the above case, the pressure recorder hooked to the high side of the system. The amperage recorder should be hooked to the condenser fan motor power supply. Each time the compressor cycles on, the pressure is recorded on a chart similar to those used in the medical profession. Everytime the compressor cycles so should the condenser fan motor. The defective motor will stop operating and thus the amperage will either fall to zero or climb to locked rotor and fall. This would be followed by the high pressure side of the system starting to elevate. This is truly the only way to solve a situation such as this.

CONDENSER UNIT AND EVAPORATOR SECTION OPERATING

■ There are times when everything appears to be operating according to all specifications, yet, insufficient cooling is the complaint.

Heat strips energizing with cooling—
Check the heat relay in the air handler and see if there is 24 volts being applied to the holding coil.

The resistive heaters would counteract the cooling effect. Sometimes the contacts of a heat relay weld in the closed position, causing this. If 24 volts are found at the relay when the thermostat mode is in cooling, you must check the system circuitry. Disconnecting for the summer is alright; however, you will have to isolate the problem sooner or later. The source should be found in the evaporator section. It is either receiving its voltage from the thermostat or the defrost device in the condensing unit, if it is a heat pump. Sub-bases are sometimes the blame. They distort on an uneven wall surface causing them to crack or short circuit.

NOT COOLING ENOUGH, ICE ON SUCTION LINE

■ Turn unit off immediately so it can defrost.

Insufficent air passing through evaporator coil—
Check for restriction, dirty evaporator coil, or filter.

■ After the unit is defrosted, and this might be the following day, place your charging manifold and gauges on the unit. Check the lowside pressure carefully. If there is a low refrigerant charge, the evaporator will start icing from the metering device back to the compressor. This of course happens over a period of time.

■ A defective metering device can cause the same problem. If insufficient refrigerant is being supplied to the coil icing could occur.

■ Restricted liquid line drier might also cause icing.

■ Another possibility is the air supply being throttled down by the room dampers located behind the supply air grill. There are times, people enter a room and feel cold. They throttle the damper and forget about it. If enough air is throttled, the unit could ice.

I would like to mention here, that in the case of glass fiber ductwork, ducts in the attic can rupture spilling cold air into the attic instead of the desired space. If you come across an extreme case when all the dampers are closed, an examination of the ductwork should be made. Even on some commercial air conditioning systems, a so-called dump damper is used to relieve supply duct pressure. It opens when pressure exceeds the design pressure of the ducting. This applies to metal ductwork as well as glass fiber. A couple of staples here and there with a little tape can be a big money saver and a more efficient unit operation.

Intermittent operation of the evaporator fan would cause the icing. The same type of testing is required in this case as with the outside fan motor.

CONDENSING UNIT OPERATING, EVAPORATOR OPERATING, INSUFFICIENT COOLING

■ Place manifold and gauges on condensing unit.

Inefficient compressor—

> Low-side pressure will be high with the high side having a lower than usual operating pressure.

■ Place clamp-on type ammeter on compressor common wire—

> The amperage will be extremely low accompanied by lower than normal operating pressures.

With this condition, some cool air is felt in the conditioned space. Complaint will be either high cost of operation and/or unit is not cooling enough on hot days. The reason is a structure requiring a three-ton unit is being cooled by a two-ton unit due to the inefficient compressor.

BLOWN FUSE OR TRIPPED CIRCUIT BREAKER

■ Use ohmmeter first to check unit continuity ground.

In checking for a ground in either section, power supply disconnects should be open. This prevents any accident from occurring due to a feed back of power. There are times when only one fuse opens letting electricity flow through the good fuse. With disconnect open, check from the load side of the circuit being tested. In the case of the condensing unit, check the compressor contactor. If a ground is indicated, all of the circuits must be isolated including the compressor, condenser fan motor, crankcase heater, and any other electrical device that may be wired to the unit. This also applies to the evaporator section when it is tested.

At times, a ground can develop in the supply circuit itself that would cause the circuit to open. In cases like these, your responsibility is over. The electrician should be notified that you are to be informed when the power supply circuit is repaired.

The third reason for power supply to open is an overloaded circuit. The demand exceeds the rating of the protection device.

UNIT STOPS OPERATION, OPERATING WHEN TECHNICIAN ARRIVES

This can present a problem due to the automatic reset controls found on most residential and light commercial units. During a malfunction, a control opens and stops the unit's operation. By the time you drive across town, the safety control has time to reset.

Flagging is done. Flags are sold at most supply houses. Let me explain to those that might never have used them. The flag is designed to conduct very small amounts of electricity. It will open like a fuse if a high ampacity load is placed across it. The flags are wired across every control in parallel. In the event the control opens, the flag will open and indicate. I've found that small ampacity type automotive fuses and holders work as well and are less costly. When flagging a unit, don't place flags on controls that are supposed to open such as the low pressure control of a pump-down system. Only flag the actual safety controls. With the use of this method, you at least find out what part of the system is the problem area.

271

CLEAN-UP OF COMPRESSOR BURN-OUT

Some might ask why I'm placing this procedure in the troubleshooting chapter of the book. Believe me when I say, much trouble can develop when the proper clean-up procedure is not followed. The sad part of it, is that the technician that replaced the first compressor is no longer around to tell you that he didn't follow procedure.

The first thing you must establish in changing a compressor was the nature of the failure. Of course if it was a mechanical problem such as a broken connecting rod, the changing of the compressor is relatively simple. In the case of an electrical failure, the clean-up is rather specific. Many times a service company can't seem to keep a compressor operational in a certain installation. The reason is the improper procedure was followed.

The severity of the burn-out is apparent sometimes when the refrigerant lines are bled. The stench of burned refrigeration oil is unmistakably pungent. Exercise extreme caution in regard to the exhausting vapor; keep if from your skin and eyes. In Chapter 14 you were told about the acid formation in refrigeration systems. Always assume the oil has acidity, and you won't get into any trouble.

When the system pressure is equalized with the atmosphere, begin taking the compressor loose from its mountings and piping. Assume nothing, spill some oil from the old compressor into a clean jar. Perform an acid test with the acid kit. The directions come in each kit. Even when there is not a foul smell to the refrigeration oil, acid still could be present in the system. If this is not done and acid is present, the insulation on the new compressor's winding will be attacked and destroyed causing electrical failure. The time between the installation and the failure is governed by the concentration of acid in the oil.

The number one thing to do is rid the system of the acid. On most commercial units using a semi-hermetic compressor changing oil is comparatively easy. There are two schools of thought about acid clean-up in a system. One is to change the suction line filter driers often, and the other is change the oil often. I've found that changing the oil and the filters is the only way to do a thorough job. Oil acts like a sponge soaking up moisture and acid. It might take several oil changes with several hours of compressor running before all traces of acid are removed.

In some commercial units, when the evaporator coil is lower than the condensing unit, it helps to place a drain valve in the lowest point of the coil. This valve is used to release any oil that might be trapped in the evaporator. With the new compressor in operation, the oil level must be checked and brought to the proper level. This is done on most semi-hermetic compressors with the crankcase sight glass.

When working on a hermetic compressor installed in a residence, the clean-up procedure is somewhat different. All manufacturers of compressors place an oil charge in the compressor when it is shipped from the factory. The compressor you are to change also has a factory oil charge in it. Oil amounts measured in ounces is not very much, yet a few more ounces than needed can create problems in a compressor. For this reason, care must be taken to measure the amount of oil reclaimed during the compressor change out.

I carried a kitchen measuring cup with a 32-ounce capacity. When the compressor was removed from the cabinet of the unit, I would pour the oil from the compressor into the measuring cup. If a compressor had a 14-ounce capacity, and only eight ounces were in the cup, six ounces were still in the system. With the use of nitrogen enough pressure can be placed into and through the piping to remove the six ounces. R-12 refrigerant can also be used as a flushing agent to remove contaminated oil from the system.

After all of that, install a suction line drier filter. In most cases, residential units were not equipped with them. Some have liquid line driers. If the unit you are working on has a liquid line drier on it, change it and then install a suction line drier. I also recommend the use of flare fittings on the suction drier, if possible. The suction line drier should be changed after running in the unit for a few days. There should be about 40 hours running time on the filter. Of course this is only an average. Clean-up of a real bad burn requires that the filter drier be changed after several hours of initial start-up of the new compressor.

It is difficult to get an oil sample from a small hermetic residential unit. For this reason, make your clean-up a proper one. Oil can be felt in the hot gas of the discharge line. This oil should only present itself in a vaporized mist. If it is experienced as oil droplets, the system has an overcharge of oil in it. This is a situation that can cause problems with the compressor, depending upon the amount of the excessive oil. The only way to remove some of it is by letting it escape from the discharge line. Of course this method requires you to top off the charge before you leave the unit.

Acid test kits are a valuable tool and not used enough by many technicians. In many commercial preventive maintenance programs that you implement, it should become standard procedure to take an acid test of a unit once a year. In many cases, a compressor could be saved by an oil change. When acid starts to form, but is removed before attacking the motor windings of the compressor, it does no harm. Years ago, an acid sight glass was used on large commercial units. This sight glass had a fine piece of copper placed where the moisture indicator is. When the service technician checked the refrigerant level in the sight glass, he was also able to see if any deterioration was taking place with the fine copper wire. If so, it alerted him to start a neutralization program. He immediately instituted oil changing keeping a constant eye on the acid sight glass to see if it progressed or ceased. I've not seen this type sight glass in parts houses for a long time. Stopping things before they happen is a big part of your job. A few drops of grease or oil in a motor bearing stops a growl, and prolongs the life of the motor.

NOISE COMPLAINT

This, in my opinion, is one of the most difficult complaints to have. People's ears are different, and different sounds and frequencies can only be heard by some. I'm not joking about it either. I can recall when condenser fan blades were made smaller with more pitch to the blade and turning at a faster speed; perhaps to keep the unit size compact. Those in the field accepted the noise as a normal operating condition. People with sensitive ears raised havoc with their complaining. Perhaps some can remember when the Carrier Corporation first introduced their 38EN models. They became a serviceman's nightmare. Realizing they had a problem, Carrier redesigned them.

If at all possible, never install the condensing unit under the window of the homeowner's bedroom. During the day sounds seem to blend. At night, they amplify. When chasing a noise complaint, a long screwdriver is desirable. By placing the metal blade against a suspected source of noise, and the handle against your ear, sounds will isolate themselves. There are many sources of noise in the condensing unit.

Internal-Mounted Compressors

The internal-mounted compressors can become noisy if one of the internal mounting springs either breaks or weakens. The unit might continue to operate efficiently with the exception of the noise. This noise is very audible. Sometimes, shifting the compressor on its mounts to change the angle of support changes the frequency of the noise to a more acceptable level. A resilient (rubber) mount under the metal foot of the compressor can be collapsed causing the metal foot of the compressor to come into metal contact of the cabinet. Some compressors have external spring mounts. The same thing applies about the resiliency of the external springs. Weather can cause them to rust and collapse. If replacement is needed, it must be for that specific tonnage compressor. All the springs might look alike, but they are rated to support a specific weight. If an overload is placed on the spring, it collapses, and it would be like having no springs at all. They are color coded or tagged for the size compressor they will support.

Hot Gas (Discharge Line) Pulsations

Pulsations from the discharge line are sometimes transmitted through the refrigerant piping into the structure. Installation of a muffler, placed as close to the compressor as possible, will lessen this condition. Check the refrigerant piping to see if it is making mechanical contact with the structure and setting up a harmonic frequency noise. Rigid pipes sometimes transmit vibration from anything within the condenser cabinet. In cases such as these you might have to install a vibration eliminator as shown in Chapter 20. The suction line, being more rigid, will probably be the one causing the problem.

Noise From Cabinet on Concrete Slab

Many condensing units are installed on a concrete slab. In many areas of the country, the slab is required by the building code. The unit should be insulated from the concrete slab by one of several means. Special rubber padding is used. Some padding has cork inside of it to add to its vibration absorption quality. The cost of this material is worth it to solve the noise problem. The pad should be cut into pieces of about four square inches. A square should be placed under each corner of the condensing unit. Not only does this method help in the solution to noise, it raises the unit high enough to allow air ventilation under the unit. This ventilation will add to the life of the cabinet by slowing the rusting process.

Motor Bearing Noise

If this noise is attended to in the very beginning, there is a good chance to stop the noise and save the motor's life at the same time.

Motors are mounted in the condensing unit and are subjected to a lot of water and moisture. It is for this reason, these motors especially should be lubricated periodically without fail. In many commercial installations, a preventive maintenance log is taped to the cabinet of the evaporator unit. On the log, the dates of servicing along with parts used and exactly what was done is recorded. This not only helps you to remember, but any other service technician that follows. For instance if you grease a motor, the date should be recorded, in that way the motor will not be over lubed.

Fan Blades

The majority of fan blades are made of either steel or aluminum. Through the years, weather causes a deterioration in both. If a fan balance weight falls off the fan, a heavy vibration begins. If a blade loses part of its surface due to deterioration, the fan will become unbalanced and cause a vibration. In either case, the fan blade should be replaced before damage to the drive motor bearings takes place. The fan shaft of the motor should be covered with a waterproof grease each time it is serviced in order to allow easy disassembly of the fan blade from the shaft. In many cases where this step is not taken during the preventive maintenance inspections, fusion of the fan shaft to the fan hub will result making it impossible to separate. In this case, you now not only need the blade, but also an expensive motor.

Reversing Valve Noise

Those people that never owned a heat pump before will complain of the swishing sound that the valve will make when going from cooling mode to heating mode. Remember this is done automatically when the unit is in the heating mode and it goes into the defrost mode. Some customers might call and report smoke coming from their condensing unit. You should inform them that this mist or vapor is normal when the ice is being melted from the coil.

Sometimes the solenoid coil mounted on the pilot valve of the reversing valve will hum or buzz. Check for tightness and fit around the stem of the pilot valve. Sometimes there might be a hum developing from one of the relays or contactors. A good non-conductive spray to lubricate, clean, and absorb moisture should be used. A set of movable contacts in the contactor could be causing this due to carbonization taking place each time it opens and closes.

Evaporator Noise

There really isn't much in the evaporator section that would cause a noise complaint. Relays or contactors might hum, if so, spray them with the same product you used in the condenser section. Lubricate fan motor bearings and belt, if belt driven. Wind noise in ducts may be due to restrictions.

ODORS

This complaint can be a problem at times. The evaporator must be kept clean along with the condensate pan. Certain strains of algae can create odors within the evaporator section. This type of condition can also cause spore formations similar to a mildew to form in the ductwork. Commercial products sold at the supply houses are in liquid, tablet, and gel form.

There is nothing more threatening or frightening than the smell of fire. This odor can be created by the windings of an electric motor overheating. The insulation within the motor, when hot, creates the pungent odor. This condition should be checked to make sure it doesn't cause a fire. Smoldering timber in the attic may be caused by the arcing of a disconnect mounted on a rafter.

Those installations that have the air handler in the garage area of the structure must be checked very carefully. The unit will suck in fumes and vapors from many things kept in the garage, such as lawn mowers stored with gasoline in their tanks, lawn fertilizers, and poison sprays. It can be a time bomb waiting to explode. Remember our craft is one that revolves around environmental control of the space used by humans. Many out there can be killing themselves in their own homes and not even know about it. All they can say is they have learned to live with that slightly displeasing smell. It's your job to explain it.

CHAPTER 32

Automotive
Air Conditioning

There are many versions of mobile air conditioning and refrigeration equipment. In this chapter, automotive air conditioning will be discussed. The air conditioning systems that you have covered up to now have an electric motor that drives the compressor. In automotive air conditioning, the internal combustion engine that drives the vehicle supplies the power to drive the compressor. This is done with a series of drive pulleys and belts. This type of system is used in automobiles, vans, and trucks.

There are other types of refrigeration and air conditioning systems that will not be covered in this book. Systems, such as those found in refrigerated trucks, use a generator to produce ac power to drive an ordinary condensing section.

Opening the hood of a vehicle and looking inside, the air conditioning system is not too hard to locate. In Fig. 32-1, the automotive-type, air conditioning compressor is shown. Notice it has two service ports like any other compressor. In automotive systems, vibrations are a constant source of refrigerant leaks; it is for that reason rubber pressure hoses are commonly used. They help absorb much of the engine and road vibrations. Also note the arrangement of the pulley on the front of the compressor. In most systems, an electro-magnetic clutch is used to cycle the compressor on and off. Most of these systems don't adhere to the rule of thumb when it comes to their tonnage. Using the rule of one ton for every 400 square feet would not accomplish the job. Due to the extreme heat

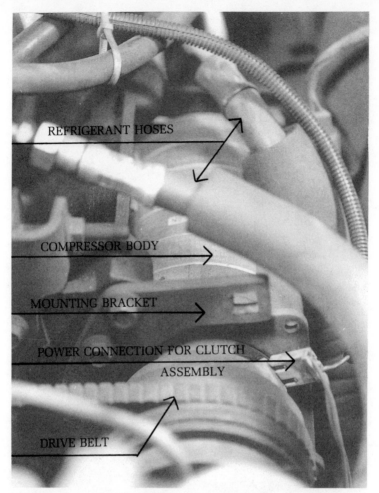

REFRIGERANT HOSES

COMPRESSOR BODY

MOUNTING BRACKET

POWER CONNECTION FOR CLUTCH
ASSEMBLY

DRIVE BELT

Fig. 32-1. Automotive-type, air conditioning compressor.

loads through the glass surfaces and lack of insulation, it isn't uncommon to find a one and a half ton unit in a vehicle. This compressor is designed with extreme rough service factor. The speed of this compressor can go from zero to 3500 rpm in a split second. It stops and goes while the engine is constantly changing speed. The compressor has a variable output rating, with a minimum and maximum, depending upon the speed range.

The clutch assembly is controlled by the on/off switch and a thermostat. When the thermostat is satisfied, the clutch de-energizes and the compressor stops pumping. Figure 32-2 shows the clutch

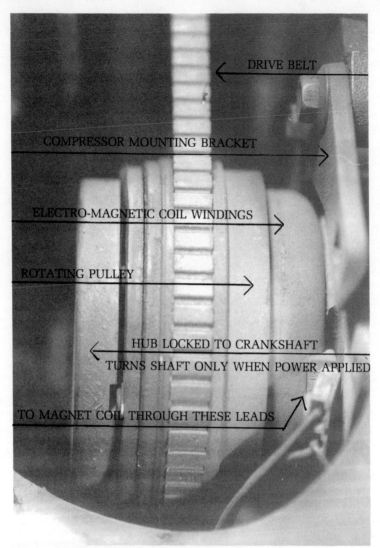

DRIVE BELT

COMPRESSOR MOUNTING BRACKET

ELECTRO-MAGNETIC COIL WINDINGS

ROTATING PULLEY

HUB LOCKED TO CRANKSHAFT
TURNS SHAFT ONLY WHEN POWER APPLIED

TO MAGNET COIL THROUGH THESE LEADS

Fig. 32-2. Clutch assembly.

assembly. This clutch can have a single or double sheaved pulley on it.

The discharge hose from the compressor leads to the receiver filter drier assembly that usually has a sight glass. Figure 32-3 is the typical receiver filter drier assembly. The hose leaving the receiver goes to the condenser coil.

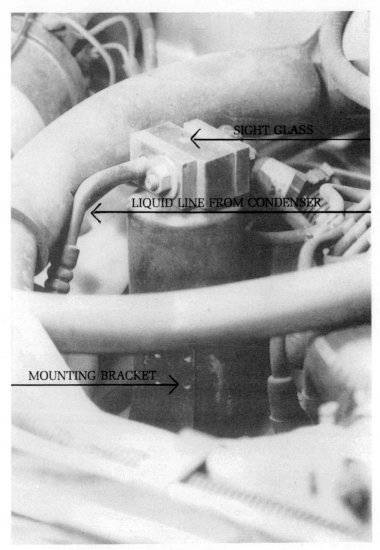

Fig. 32-3. Receiver with filter drier.

Figure 32-4 is a typical automotive air conditioning condenser coil. The condenser coil on most automobiles is placed directly in front of the radiator, and uses the regular fan to draw air across it when standing still. When the vehicle is moving ram air passes over the condenser coil and condenses the refrigerant. The condenser on the automobile should be washed frequently. Road dirt and dead

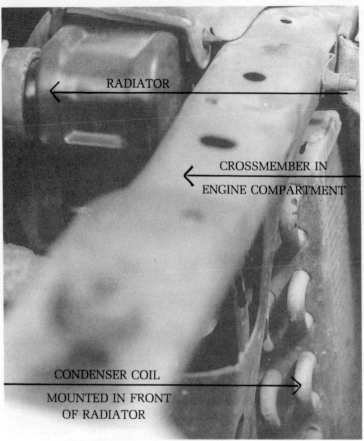

RADIATOR

CROSSMEMBER IN
ENGINE COMPARTMENT

CONDENSER COIL
MOUNTED IN FRONT
OF RADIATOR

Fig. 32-4. Automotive condenser coil.

insects cause a restriction in the air flow, for that reason, a water hose is used to flush the condenser clean. The hose leaving the condenser is the liquid line, and it leads to the temperature expansion valve. You can see in Fig. 32-5 the expansion valve is very similar to those you have been working with. Before going any further, look at Fig. 32-6, this shows a pressure control. These controls are wired in series through the clutch and will prevent the compressor from operating when pressure is too high or low. With the use of these controls, manufacturers of the compressors were able to cut down the amount of warranty compressor replacements. It is for this reason that you should not jump out any controls. If a control is found defective or there is a situation causing the control to open, make the necessary repair, and only then let the unit run.

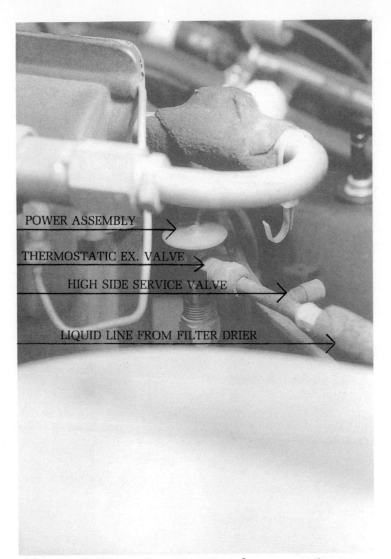

POWER ASSEMBLY

THERMOSTATIC EX. VALVE

HIGH SIDE SERVICE VALVE

LIQUID LINE FROM FILTER DRIER

Fig. 32-5. Thermostatic expansion valve, automotive type.

The evaporator coil and fan drive motor appear in different locations. Some are mounted in the engine compartment and in others they are mounted under the dash board. Unfortunately to service these in most cases involves the expertise of the technician that chose automobile air conditioning. It requires many specialty tools that the average technician would not have in his tool box. Leave

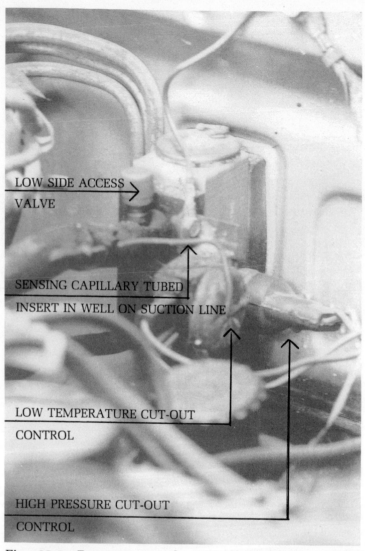

LOW SIDE ACCESS VALVE

SENSING CAPILLARY TUBED INSERT IN WELL ON SUCTION LINE

LOW TEMPERATURE CUT-OUT CONTROL

HIGH PRESSURE CUT-OUT CONTROL

Fig. 32-6. Pressure control stops the compressor from operating if pressure is too high or too low.

major repairs concerning the evaporator to the automotive air conditioning service technician.

Compressor drive belts are made specially for automobiles as described in Chapter 24. When the drive belt becomes glazed, it will cause a squeal. This is due to its loss of traction. Overtightening the drive belt to remove the noise will only place excessive loads on the clutch bearings and any other accessory driven by the same belt, such as the alternator or power steering pump. An overtightened belt can ruin an alternator, power steering pump, or water pump in a very short time. A new belt will make the compressor more efficient by not slipping on the pulley of the clutch. Another thing is the heat that is generated by the drive belt slipping can cause the lubricating grease of the clutch bearing to melt and migrate from the bearing causing failure. You have no way to use a clamp-on ammeter to see if the belt is too tight; it is adjusted with feel and a little slack.

Caution should be exercised when charging an automotive air conditioning system. A clear face mask should be used to protect eyes and face. The charge in an automobile system ranges from one-and-a-half pounds to three pounds of refrigerant. The reason for extra care is due to moving fans, belts, and rubber refrigeration hoses that can rupture without warning. In servicing the system, refrigerant can be handled in one pound containers. If you have a system that is completely out of refrigerant, it should be charged with a weighing cylinder, sometimes called the dial-a-charge. For topping the charge, clean off the sight glass at the receiver filter drier. Watch the sight glass while charging. As soon as the flashing in the glass stops, stop charging. A window fan or floor fan aimed at the condenser coil will simulate ram air during driving conditions.

Refrigerant 12 is used in the majority of automotive air-conditioning systems. Head pressure in these systems will be much higher than a regular R-12 system. Pressures will normally range from 165 to 300 pounds. In other applications these pressures would be unacceptable. That is another reason the compressors are specially designed units to withstand a lot of abuse. The high pressure in this system makes the high pressure control an absolute must in the control system. If it is bypassed or jumped out, you can see that the pressure would continue to rise until the weak link of the system would rupture.

The charge in the system is very critical. For this reason each time a service technician places his service manifold and gauges onto a Schraeder service valve, a couple of ounces is lost in the charge. In a system that is using seven to ten pounds a couple of ounces lost is hardly noticeable. With a system that might only have one pound, eleven ounces in it, the loss of an ounce can be critical and might cause problems such as evaporator coil icing or insufficient cooling. Due to the small amount of refrigerant used in the system, a complete charge should be measured in with the dial-a-charge. This assures a longer life in the unit by preventing over charge or under charge.

Condensate drain usually drips from the hose connected to the condensate drain pan. A puddle of water will collect under the vehicle if it stays in one place long enough. At times, it is possible for this drain to become restricted and overflowing will occur. This can cause a major problem by soaking the carpeting. Unlike the evaporator section located in a structure, no provisions are made in the automobile to have a filter on the return air to keep the evaporator clean. The same thing applies to cleaning the condensate pan. These are not common occurrences in automotive air conditioning servicing. If drain blockage does occur, try to clear it from the outlet side of the condensate drain line where it exits the vehicle.

CHAPTER 33

Heat Reclaim

When the country went through the energy crisis several years ago, many new products were introduced to the market that were designed to save energy. In the air conditioning industry a very simple idea was put into action, and people were producing hot water with their air conditioning systems.

In the majority of residential homes, air cooled condensing units are used. The heat from within the structure was transferred from the inside to the outside ambient. The heat was rejected into the atmosphere. Someone thought that this was a waste of energy. That person sat down and developed an idea to re-use the heat of rejection.

The discharge line from the compressor is routed to a tube-and-tube heat exchanger. Figure 33-1 shows how the discharge line enters and exits the heat exchanger. The return line to the compressor then hooks up to the air-cooled condenser coil as it usually would. A small water pump is located in the heat exchanger cabinet. Two water pipes from the structure's hot water heater enter the heat exchanger and leaves it. The small circulating pump circulates the water in a loop through the water heater. Two controls located in the cabinet are wired in series to the pump. Both controls must be closed in order for the pump to operate. One control senses the discharge gas temperature. If it is not hot enough, the pump will not circulate the water. The other control senses the temperature of the return water from the hot water heater. When this temperature reaches a pre-set limit, the control opens. This

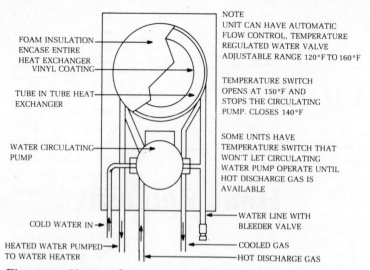

NOTE
UNIT CAN HAVE AUTOMATIC
FLOW CONTROL, TEMPERATURE
REGULATED WATER VALVE
ADJUSTABLE RANGE 120°F TO 160°F

FOAM INSULATION
ENCASE ENTIRE
HEAT EXCHANGER
VINYL COATING

TEMPERATURE SWITCH
OPENS AT 150°F AND
STOPS THE CIRCULATING
PUMP. CLOSES 140°F

TUBE IN TUBE HEAT
EXCHANGER

SOME UNITS HAVE
TEMPERATURE SWITCH THAT
WON'T LET CIRCULATING
WATER PUMP OPERATE UNTIL
HOT DISCHARGE GAS IS
AVAILABLE

WATER CIRCULATING
PUMP

WATER LINE WITH
BLEEDER VALVE

COLD WATER IN

HEATED WATER PUMPED
TO WATER HEATER

COOLED GAS

HOT DISCHARGE GAS

Fig. 33-1. Heat reclaim unit used for producing hot water.

prevents the water in the hot water heater from reaching high enough
temperatures to open the pressure relief valve.

The thing to remember is that the air conditioning system has
to be operational in order to heat the water. Many people have their
air conditioning units turned up very high on the thermostat in order to reduce their monthly power bills. This heat reclaim system
will not work unless the air conditioning system is operating normally.
This system has been used in many commercial establishments with
great success. In fact a payback has been figured that in most cases
the initial cost of the unit is returned in savings over a three-year
period.

Some hotels, motels, and restaurants have been using this type
of equipment with great success. Again I would like to say, that this
piece of equipment works well and will pay for itself if used in an
area where air conditioning seasons are not too short. This would
not be a practical idea in Hudson Bay Canada.

In the heating mode during the cold weather, the system is not
efficient. The heat that is generated by the heat pump is needed
for the comfort of the people in the conditioned space. The air
entering the structure would be too cold if it transferred its heat
to the water. For this reason, the water heater will have to be
supplied with its own energy source to heat the water when the
weather turns cold. In the summer, the savings could be high enough
to defray the cost of installation.

Remember, as in the case of the tube-and-tube condenser found in water-cooled air conditioning systems, a possibility for chemical build up inside the tubing can cause an insulation effect inhibiting good heat transfer. This means that there might have to be a periodic acid cleaning of the heat exchanger, depending on what area and what hardness the water is.

The commercial application of this type of water heating is very effective. Usually where there is a lot of motion in a conditioned space, even during the cold season, people get hot and the air conditioning is required. Recall the earlier example of the banquet hall. The same thing applies here. The dishwasher is supplied with free hot water as long as the air conditioning units are operating.

A pressure relief valve is supplied on the water circuitry in the heat reclaim unit. This relieves the water pressure outside the structure in case there is a malfunction in the temperature control. These units are usually mounted close to the condensing unit. It is easier and a better condition to extend water piping rather than refrigeration piping.

Electrical power is taken from the main supply to the condensing unit. The amperage for the small water pump is less than one amp in most cases. There are two small fuses in the unit to protect the pump against over current. It is suggested that the line voltage power be taken from the load side of the compressor contactor. The reason for this is there is no reason to keep the heat reclaim unit operating if the compressor is not. Wiring the unit in this manner guarantees the circulating pump will only operate when a heat source is operating, namely the compressor.

A toggle switch is located in the heat reclaim unit to open the line voltage circuit. A drain cock is also located in the heat reclaim unit. This is used to drain the unit to prevent freezing if it is not used in the winter season.

CHAPTER 34

Refrigeration

In this chapter, refrigeration will be discussed in two sections, light commercial and domestic. The refrigeration theory that was discussed earlier in the book applies in all applications of refrigeration. Transferring heat from an area where it is not desirable to an area where it is not objectionable.

REMOTE SYSTEMS

This system is similar in a way to the split air conditioning system. The evaporator is located usually within the structure and the condensing unit is outside in the ambient. This type of installation can apply to reach-in units as well as walk-in units. The advantage to this type of installation is the heat rejection is kept outside the structure along with the noise of operation. The two sections are connected with piping. In light commercial applications, most units you will be working on will be either medium or low temperature range equipment.

Refrigeration is used for the preservation of a commodity. This fact must be totally understood before entering into the refrigeration field seriously. You will learn that refrigeration is used from keeping liquid oxygen to keeping garbage from deteriorating and causing possible disease.

With the span of applications that refrigeration covers, it is an absolute necessity that you know what the unit you are working on is supposed to do. Most refrigeration systems were designed on a drawing board to perform a specific job; trying to change that only ends up with a lot of aggravation and much frustration for both you and the customer.

Walk-In Cooler

This unit can be commonly seen in most convenience stores, supermarkets, and restaurants, to name a few. The average that you'll see in this application will range in size from a six by six foot with an eight-foot ceiling to a 15 foot by 20 foot with an eight-foot ceiling. This is not to say that they are not built in larger sizes, they are, but they will not be covered in this book. The units to be discussed here will be rated to be driven by a condensing unit from one to five hp. See Fig. 34-1 for a typical restaurant application. This particular unit is a pre-fabricated system that is put together like a child's toy on Christmas. This method has allowed a lot of do-it yourselfers an opportunity to install this equipment. With the use of pre-charged refrigerant lines, a novice can install a unit such as this one.

In the conditioned space the temperature of the product could be anywhere from 42 degrees F. to 32 degrees F. In some cases this can become a difficult task. If the customer wants to have the coldest beer in town, he certainly isn't going to have the best looking

Fig. 34-1. Walk-in freezer or refrigerator.

produce. The low temperature would remove too much humidity from the produce. This is one of the reasons, you have to watch for these things. An evaporator coil is hung in the walk-in cooler. The evaporator could have fans that cause air movement over the coil, or it could be the convectional type of coil. This coil has widely spaced fins and usually drops condensate on the floor of the walk-in, leaving a high humidity factor. If the humidity is not needed, condensate pans can be used to remove the condensate water from the unit. Many seafood coolers operate that way. The temperature is controlled with the use of a control that is mounted with its sensing device in the return air of the evaporator coil. Figure 34-2 shows the typical control. This control might be wired to only cycle the machine on and off. It is recommended that a liquid-line solenoid be installed in the system. The temperature control would only activate the solenoid, and the unit would cycle on and off with the use of a low-pressure control located in the condensing section.

Most refrigeration coils are designed to operate with a TEX valve set with a superheat of 10 degrees. This would mean that the

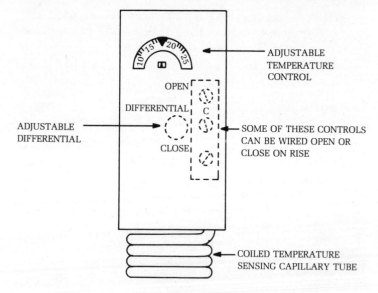

NOTE
TEMPERATURE CONTROLS USED FOR REFRIGERATION
APPLICATION SHOULD BE INSTALLED TO BE
ABLE TO SENSE RETURN AIR TEMPERATURE AT THE
EVAPORATOR COIL

Fig. 34-2. Temperature control, commercial type.

coil would lead the air passing over it by 10 degrees. For instance, if the coil temperature (refrigerant temperature) is 32 degrees F., the supply air in the refrigerated space would be 42 degrees. If the condition is held at a constant, all of the commodities in the refrigerated space would attain the same 42 degrees as the air. In reality, the temperature keeps dropping and the coil temperature is always 10 degrees lower than the air passing over it. For this reason, it is suggested that there also be a provision for a time defrost in cases such as this one. It is not necessary to have this defrost device if the unit is used for items that don't require severe cold, such as leaf vegetables. Back to the beverage temperature of 34 degrees, you can see that an evaporator coil might have to achieve a temperature around 24 degrees F. If there wasn't any warm moist air being introduced to the refrigerated space, the defrost device would not be necessary; however, people are constantly opening and closing the doors of the cooler. Each time this happens, the warm, moisture-laden air is pulled to the cold evaporator coil where the moisture turns into frost. This condition doesn't require any exotic, sophisticated controls to be installed. An inexpensive 24-hour time clock is used to cycle the liquid-line pump-down solenoid.

A condensate drain line is used for those applications that do not desire any humidity. Fan switches or wire caps (plugs into receptacles) are usually used to disconnect the power supply to the fans in the evaporator section. A floor drain is an asset to the installation for cleaning purposes. A water hose makes cleaning the evaporator coil a quick task.

Before going any further, let me mention defrosting the walk-in cooler. I suggest as a starting point, four defrosts during a 24-hour period with a period of 20 to 30 minutes. Remember that the defrost cycling will depend on the individual application of the unit. Two exact units will require two different defrost programs due to the amount of traffic and the people that are opening and closing the doors.

Be observant as to the way the customer is stacking products in the walk-in cooler. Blocking of sufficient air circulation could cause rapid icing and inefficient operation.

Condensing Section

This unit can be located either in the back room, in the back yard, or up on the roof. To me the roof is the best place so the unit is not vandalized nor has things stacked around it that chokes off

Fig. 34-3. Typical refrigeration condensing unit.

the air it needs to perform properly. Figure 34-3 illustrates the typical condensing unit.

Most of the controls used in this type of system are almost the same as those used in high temperature systems, (air conditioning). Depending once again on the geographical location of equipment, controls will vary. In many of the older walk-in coolers, the only control that is used for regulating the temperature of the conditioned space is the low pressure control.

The cut-in pressure would be the high temperature event in the walk-in cooler. The cut-out pressure would be the low temperature event. All compressors must run a minimum amount of time in order to bring the oil back to the crankcase. If this is not done, eventual failure will occur. Another rule-of-thumb is to allow a compressor to have at least a five-minute run time, whenever it starts. To do this with the low-pressure control, the differential adjustment must be made as to allow for this operating time. Most of the time, a differential of 10 psi is adequate. An example of this would be, in a R-12 system that has a cut-in of 38.8 psig and a cut-out of 30.1 psig. The results of this in the units operation are as follows.

      38.8 psig   temperature of 42 degrees F.
      30.1 psig   temperature of 32 degrees F.

The differential you can see is 10 degrees. This operating time from 42 to 32 degrees in the cooler would achieve an average temperature of 37 degrees. This is the method you use to control the temperature of the cooler with the use of a low-pressure control only.

One of the biggest problems with this type of control is cold weather. You can easily see if the condensing unit is in the cold ambient of perhaps 20 degrees, the customer's products might need refrigeration; however, the cold air is causing a slight problem in the expansion of the refrigerant that would allow it to reach the necessary pressure of 38.8 psig to cycle the equipment on. This is why the pump-down system is more efficient in certain applications.

Fan cycling switches should be used to maintain proper condensing temperature. This can be achieved by an inexpensive bi-metal disc switch or a reverse-acting, high-pressure control. Crankcase heaters are important, especially during cold weather operation in northern latitudes. Face dampers over the condenser coil might be needed if there is a prevailing wind. These dampers can be operated by the refrigerant pressure or electrically. Many light commercial installations use a small refrigeration house (enclosure) on the roof to shield the multiple condensing units from the weather. Dampers and fans control the amount of ambient air that flows over the condensers.

Walk-in Freezers

Freezers are more difficult than the walk-in coolers. There are many more components involved to make it operate properly. In the majority of walk-in coolers, R-12 refrigerant is used. In freezers, the two basic refrigerants used are R-12 and R-502. I'll get into that deeper in a little while.

Inside the walk-in freezer, the apparent thing you first see is the ice formed on the coil. The evaporator is different than the cooler. The major difference being a defrost system that will quickly and efficiently remove the ice from the evaporator coil so that air might once again flow through it. This is accomplished in many ways. Figure 34-4 shows a wiring schematic of a typical and very popular method of defrosting.

Resistive rods, called Calrods, are attached to the evaporator coil in different sectors. Some are on a vertical plane of the coil, and others are on a horizontal plane under the coil. The condensate pan is also equipped with a Calrod. These resistive rods create a lot of heat when electricity is passed through them. The defrost cycle

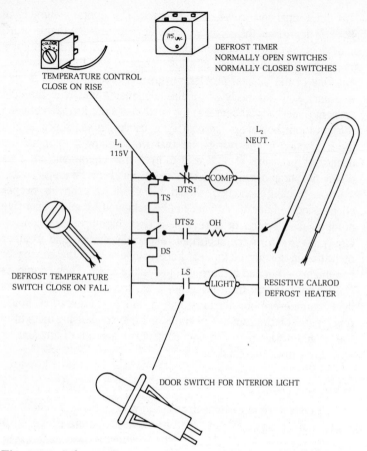

TEMPERATURE CONTROL
CLOSE ON RISE

DEFROST TIMER
NORMALLY OPEN SWITCHES
NORMALLY CLOSED SWITCHES

L_1
115V

L_2
NEUT.

DTS1

COMP

TS

DTS2   OH

DS

DEFROST TEMPERATURE
SWITCH CLOSE ON FALL

LS

LIGHT

RESISTIVE CALROD
DEFROST HEATER

DOOR SWITCH FOR INTERIOR LIGHT

Fig. 34-4. Schematic wiring diagram of defrost heater circuit
and switches.

will start with the liquid-line pump-down solenoid closing. This will
evacuate all the refrigerant from the evaporator coil. The amount
of heat that is going to be generated there would create high pressure
within the coil that is really not necessary. When all of the refrigerant
is removed, the system will stop operating. The defrost timer will
energize the defrost heaters. The condensate drain line in the freezer
should be wrapped with a pipe heater tape to prevent icing of the
drain line. This heater should be kept energized at all times. The
defrost cycle program has to be designed for each specific unit. Ice
should melt from the coil and the condensate pan; in most cases,
30 to 45 minutes will lapse before the unit is free from ice.

Defrost is terminated either by time or by pressure. When the unit is cycled on to the freeze cycle again, one thing is different. In many units, the evaporator fan motor will not operate until the temperature of the evaporator coil reaches near zero degrees F. The reason for this is not to circulate the hot air from the heaters onto the product. This would place an unnecessary heat load into the box. The fan will begin to operate when the heat is removed.

At this time, I would like you to make a mental note about the installation of dry desiccant filter driers. Never, install them in the conditioned,refrigerated space. If a filter drier must be used within the refrigerated space use a Silica-gel type. The ordinary filter drier will give up its moisture and cause icing in the expansion valve.

Another device that should be noticeable in the area of the evaporator should be a heat exchanger. In Figure 34-5 a heat exchanger is shown. The purpose of this device is to vaporize any liquid that might have passed through the coil. This will help prevent slugging. The liquid line is also sub-cooled in this manner to reduce the amount of refrigerant needed for flash gas.

Condensing Section

It is at the condensing section that you will see major differences in the equipment. Due to the temperature of the refrigerant, preventive steps must be taken to protect the compressor against injury from slugging. I'm not going to spend time explaining the accumulator. It works in the same way in refrigeration as it does with the heat pump. An oil separator is rarely seen on light commercial equipment. It is a device that removes oil from the discharge gas and returns it to the compressor crankcase.

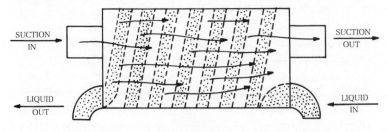

SUCTION IN   SUCTION OUT
LIQUID OUT   LIQUID IN

Fig. 34-5. Typical heat exchanger sub-cools liquid and helps to boil off any liquid refrigerant that didn't boil in the evaporator. This arrangement helps to prevent slugging.

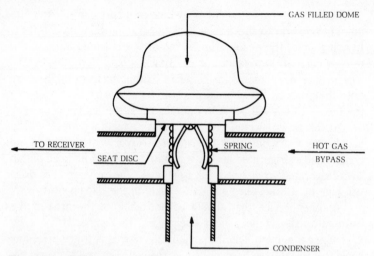

GAS FILLED DOME

TO RECEIVER

SEAT DISC

SPRING

HOT GAS
BYPASS

CONDENSER

Fig. 34-6. ORO valve (opens-on-rise-of-outlet), pressure in the receiver throttles this valve, helps maintain proper pressures in a cold ambient.

Fans have cycling devices, as described in the section on walk-in coolers. In addition to the fan cycling, many freezers use a head pressure control device. In Fig. 34-6 an open-on-rise-of-outlet valve is shown. Commonly called the ORO valve. This valve enables the compressor to pump discharge gas into the liquid receiver causing the head pressure to rise. In Fig. 34-7 you can see the routing through the ORO valve.

In most of the newer equipment being used in low temperature, the refrigerant is R-502. Compressors do not have to be as large as the ones pumping R-12. Many of the very low temperature ranges needed caused R-12 to be operated close to, or within a vacuum. This can be very critical if there are any minute leaks in the refrigerant circuitry. For this reason, R-502 is preferred.

Don't forget the expression, design factor. The design factor are those specifics that apply to the unit during its normal operation. This is very critical when servicing low temperature refrigeration equipment. Many service technicians forget design factor. When topping a charge, or completely charging a system, don't cause excess current draw by the compressor. This is very easy to do. Sight glasses can flash when a coil is heavily iced, the freezer door is open, or the unit is just going back into freeze from the defrost cycle. When a unit is at a high temperature, it is not at the design

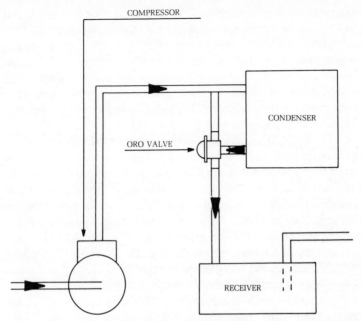

Fig. 34-7. Gas routing through the ORO valve, in normal operation.

temperature of a low temperature freezer. I can't emphasize it enough, for this is one of the main reasons for failures, overcharging a unit.

REACH-IN COOLERS

A reach-in cooler is a medium temperature appliance. It can either be a remote installation or a self-contained unit. These units are also used in convenience stores, restaurants, and supermarkets. If the unit is a self-contained appliance, the condensing unit will either be on top or under the cabinet.

These units might have either an expansion-type or a capillary-type of metering device. If a capillary-tube type of device is used, the coil temperature might be designed to operate at a low temperature, to effect quick recovery when the door is opened. A temperature control is used to cycle the unit off and on. The unit might have a coil designed for a temperature as low as zero degrees F. This is why you will sometimes find a beer cooler that begins to freeze the beer and wine. If the temperature control fails, you can have all kinds of problems. With these commercial units, you will find certain malfunctions aside from the normal ones.

Condenser coils seem to get excessively dirty when used in restaurants. Dust and dirt can be handled with a good vacuum cleaner. In the case of cooking grease, a spray of oven cleaner does a good job. Flush with a lot of water. A wet-vac is very helpful when the cleaning of this equipment is in a kitchen and it can't be removed for outside cleaning.

The next problem is leaks. Many refrigerant leaks occur in reach-in units that are placed in restaurants. The reason for this is acetic acid, that causes minute pin hole leaks in the evaporator coil. The common word that we all know is vinegar. The refrigerators that house the pickles, salads, any other condiments or preservatives fall prey to the leaks. They are usually easy to find once the evaporator is exposed for examination. The difficulty is in the disassembly of the unit. Most of the time, the mounting screws have deteriorated beyond recognition. It might be necessary to replace the evaporator coil, if there are too many holes eaten in it.

Another problem is door gaskets. These are seals that keep the cold air inside the space and the hot air outside the space. There are times that people in the kitchen don't believe this and tear gaskets from the doors. Of course, a unit can't perform as it was designed. There are times that a door is slammed against a tray or some other object that is in the way of proper closure. This results in the door either warping or being thrown out of alignment with the cabinet. If you ever need cabinet hardware or parts, get the manufacturer's name and address and possible part numbers that might be stamped on the part needed. The reason for this, is there are many vendors that manufacture these hardware parts for the refrigeration company who builds the reach-in equipment.

The reach-in refrigerator can be a combination unit that has a refrigerator on one side and a freezer in the same cabinet on the other side. These side-by-sides and over-and-under units are very popular. These combination units have two separate refrigeration systems within the single cabinet. These units either have a piped drain to them, under or behind the cabinet, or they will have a condensate evaporator pan under them. This pan has a resistive heater in it which will actually boil water in it. This adds to the humidity in the room, but in a commercial kitchen, the exhaust fans and make-up air fans change the air fast enough so this is not a problem. There are many companies that are producing this type of equipment, don't be afraid of them. They are all using the refrigeration principle which you already know. In light commercial refrigeration, many of the controls used are used in other applications so that you should be familiar with them.

Multiple Evaporators

In many cases, you might find that one condensing section is driving several reach-ins, or for that matter walk-ins. Don't get fearful of this either. This is simply another way to economize on the installation. Through the years, the owners are switching away from this type of installation due to all of the merchandise depending upon one condensing unit. This is alright if there are spare parts available in the machine room; however, more times than not, there aren't any. Many of the supermarket chains use this type of installation due to the fact they have so many refrigerated cases. Two or three horizontal or vertical cases are driven by one condensing unit. This is done with the use of temperature-controlled solenoids that were previously explained. Each case has its own control and liquid line solenoid. In some cases, a compressor will unload itself as the load is shed. When all of the cases are satisfied and the temperature control closes the liquid-line solenoid, the compressor pumps itself down and stops its operation with the use of the low-pressure control. Figure 34-8 shows a multiple evaporator refrigeration system.

If you decide to specialize in refrigeration, commercial and industrial, you will find more of a diversification as to the application of the refrigeration theory. Always come back to the basics when you feel a little confused. The refrigeration theory touches most of our lives everyday in one form or another. What did the design engineer want this unit to do? That's the first question you should ask yourself when troubleshooting.

DOMESTIC REFRIGERATOR

This one of the silent servants of the home. It, like the air conditioning system, is usually forgotten about until there is a malfunction. The majority of household refrigerators are capillary tube systems that are powered with a fractional horsepower hermetic compressor. They can also be identified as straight refrigerators that might have a small freezer compartment within for making a couple of trays of ice cubes. The majority of the newer machines have separate doors and compartments for the refrigerator section and the freezer section. These are combination boxes as are found in

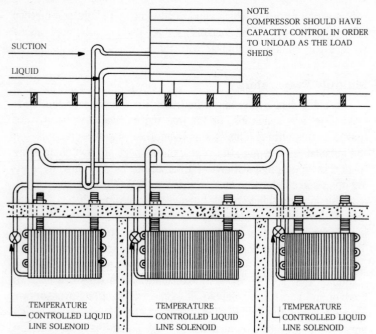

SUCTION

LIQUID

NOTE
COMPRESSOR SHOULD HAVE
CAPACITY CONTROL IN ORDER
TO UNLOAD AS THE LOAD
SHEDS

TEMPERATURE
CONTROLLED LIQUID
LINE SOLENOID

TEMPERATURE
CONTROLLED LIQUID
LINE SOLENOID

TEMPERATURE
CONTROLLED LIQUID
LINE SOLENOID

Fig. 34-8. Multiple evaporators being driven by a single condensing unit.

commercial applications. They are either over-and-under or, side-by-side. unlike some of the commercial equipment with two condensing units, the domestic unit usually has only one. The most important factor to consider here, is that there isn't as much in-and-out traffic in the refrigeration equipment in the home. Even with little children constantly going in and out of the unit, it doesn't compare to the heat load of a commercial kitchen. For that reason, domestic refrigerators run for many years trouble free.

Checking the Power Supply

In most refrigerators a light bulb is energized when the cabinet door is opened. The door jamb switch accomplishes this. If the light comes on, you are automatically informed that there is electric power to the unit. If the light doesn't come on, there might be a lack of power, or a bad light bulb. This is easy to check with your voltmeter and ohmmeter. If the solution is not as easy as plugging in the service cord, the unit will be pulled away from the wall. Take care not to mar the flooring. Placement of towels or small throw rugs under the front of the unit might help it slide. Be very careful not to damage anything hanging in the back.

Condenser Coil

On many domestic refrigerators, a convection type of condensing coil is used. It is mounted vertically to the back panel of the unit. It usually stands away from the panel by a few inches to allow the circulation of air to pass through the coil. The convection currents of air are caused by the fact that hot air rises. The refrigerant condenses here in the convection coil. It is also back here that you find all those objects that you have lost through the years. The dust and dirt will also be present so don't start blaming someone for being a poor housekeeper. Any airborne dust in the room might become attached to the coil due to the convection currents in the air.

There really isn't much that can go wrong with this part of the system. A hole can be created, but it is very unlikely. In Fig. 34-9 a picture of a typical convection coil is shown. This component should be kept as clean as possible with a brush and a vacuum cleaner. Located in the back panel there is an electrical connect compartment. In this area, a defrost timer can be seen. In some units these components might be located elsewhere; it is impossible to show every unit in a book of this type. Figure 34-10 shows the typical defrost timer. This will give you an idea of what the part looks like. There are variations, but they all basically function the same way. Looking down towards the floor, you will notice a compartment where the condensing section is located.

Condensing Section

The compressor will be the first thing you will recognize. It will be small and a fractional-hp hermetic. Be careful in touching anything. First, the unit should have been disconnected from the power supply before it was moved from the wall. Second, the compressor could be extremely hot. This can be due to,

- Insufficient refrigerant charge
- Locked rotor
  —bearings seized, lack of lubrication
  —defective starting component
  —defective windings, defective compressor
- Power fluctuation
  —compressor stalled for a second and tried to restart
  —overload cycling the unit

303

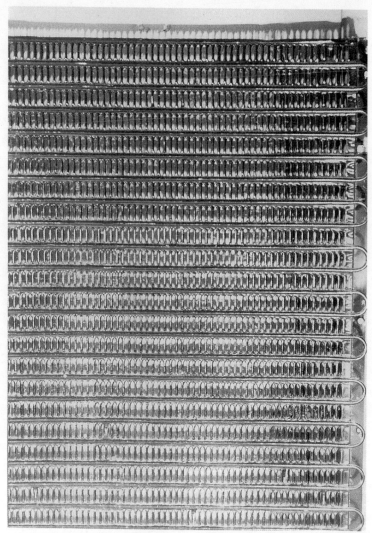

Fig. 34-9. Convection type of condenser coil located on the rear panel of most domestic refrigerators.

If the compressor is cold, it is a sign that it hasn't been operational. Check the circuitry with an ohmmeter. The small start relay can be the problem. It plugs on the compressor terminals. Some units might have a small fan in this compartment. It is used to cool the compressor as well as induce evaporation of the condensate water and the defrost water. This section also has a condensate pan. Some

Fig. 34-10. Domestic refrigerator defrost timer.

manufacturers have routed their hot gas discharge line from the compressor into this condensate pan. The hot gas helps the water, and in turn, the hot gas sub-cools before it enters the condenser coil. Everything in this section should be cleaned at this time. Make the necessary compressor tests as described in Chapter 3. After establishing the condition of the compressor, you will have to open the sealed refrigeration system. The line piercing valve can be used here. It makes an easy service port. I would rather use a Schraeder valve, hard soldered into the port where the factory placed its crimped fitting. Whatever device you choose to use, you must get into the sealed system to check the refrigerant charge. I can't impress on you enough to exercise extreme caution so as not to introduce air and moisture into the system. Before valves are opened, be sure that the service manifold and hose are purged of air. With the gauges now attached to the unit, you can see whether there is a lack of charge. If there is a leak and the charge has completely leaked out, the gauges will show zero pressure. Proceed with leak testing in the condenser compartment, if no pressure is present on the gauges.

If refrigerant pressure is present in the system, the unit might have a defective defrost timer. If this is the problem, the compressor will probably be cool to the touch for it will not have been operating. This will also be indicated with a heat load inside the unit. Things inside the storage area, should have been removed from the unit. Inside the unit, there is an evaporator fan motor and blade, an evaporator coil, a thermostat, and a thermodisc. These are the major components that are located within the refrigerated space. Of course there are ice makers and other accessory items. Although there are many manufacturers of this equipment, they all basically contain the major parts listed above.

The domestic refrigerator has the same type of defrost system as does the light commercial units. With a single evaporator coil in the average unit, the cabinet is divided into two compartments, a low-temperature and medium-temperature storage areas. The single temperature control has its sensing bulb mounted in the medium-temperature section. This temperature control cycles the compressor on and off. The cold air from the evaporator coil is blown into the freezer section first. This air is usually around zero to 10 degrees below zero. It removes heat from the freezer section, and then it flows through the refrigerator section before it returns to the evaporator coil. With this arrangement, the appliance can maintain freezer temperatures between zero and 10 degrees F. below zero, and refrigerator temperatures upwards from 32 degrees F.

Evaporator Fan and Motor

Panels will have to be removed inside the unit, in order to expose the fan motor and fan blade for servicing. Most units use the squirrel-cage type fan and a sealed bearing motor. They don't usually require service, only replacement. At times, the fan motor mounting bolts might become rusted and deteriorate to a point that the motor actually shifts its position and either become noisy or binds against the fan scroll. By removing the plastic liner, the fan section can be seen. This is done in different ways, depending upon the brand. The evaporator drain pan is piped to run the defrost water down the rear wall of the conditioned space. This water helps replace some of the moisture in the crisper drawer. The refrigeration effect causes the produce to dry out; with the drain being designed this way, some moisture is replenished helping the produce to stay fresher longer. The excess water is then routed to a drain line that takes the water into a pan beneath the unit. This pan, located in the compressor compartment, evaporates the water either by directing fast-moving air across the water or by heating and evaporating it. The heat can be supplied by discharge gas, Calrod heaters, or other heat-generating device.

Defrosting

With the evaporator coil operating at the low temperature, the

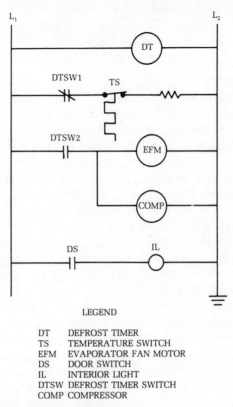

LEGEND

DT	DEFROST TIMER
TS	TEMPERATURE SWITCH
EFM	EVAPORATOR FAN MOTOR
DS	DOOR SWITCH
IL	INTERIOR LIGHT
DTSW	DEFROST TIMER SWITCH
COMP	COMPRESSOR

Fig. 34-11. Schematic of the simple defrost circuit in domestic refrigerators.

formation of ice on the coil is persistent. The more frequent the usage of the box, the heavier is the ice accumulation and the more frequent the defrost cycles. This is accomplished in a very simple way.

Mounted to the evaporator coil is a defrost heater Calrod that is very similar to those found in commercial freezers. Although this one is smaller, it works the same way. The rod is made of a low resistance alloy that generates a high temperature when an electrical current is introduced to it. In Fig. 34-11 a simple schematic diagram is shown how the components of the defrost circuit are wired.

A thermodisc is mounted to the evaporator coil near the outlet pipe. This is a bi-metal switch that opens on rise. The temperature required to close this switch depends on the specific unit. For example, if the thermodisc is designed to close at 20 degrees F. and open at 70 degrees F. the Calrod can only receive electrical power when the evaporator coil is below 20 degrees F.

Defrost Timer

The defrost timer runs continually. Contacts close at different intervals, again depending upon the manufacturer. A timer might close contacts for several seconds every 30 minutes. If the thermodisc switch is open, the defrost timer will begin another 30-minute cycle without anything happening. When the thermodisc switch closes, and the defrost timer contacts close for the several seconds, the unit will go into defrost cycle. The compressor stop operating, the evaporator fan motor stops operating, and the electrical power flow from the defrost timer switches to the thermodisc across its contacts to the Calrod defrost heater. The heater will stay energized until the thermodisc switch contacts open. At that time, the compressor and evaporator fan motor will start operating again. The Calrod heater is de-energized and cools. The unit stays in the refrigeration cycle until the thermodisc senses a defrost is needed again. An iced coil inhibits the circulation of air across the cold tubing, it is for this reason that you will notice that the spacing of the fins and tubing on freezer coils are much further apart than they are in air conditioning equipment. When the evaporator becomes completely blocked by ice, the product will begin to pick up heat. This is the reason for programmed defrost cycles.

CHAPTER 35

Tools

I've been asked many times by many technicians, "What tools should I have?" This is a very difficult question to answer due to the fact that every refrigeration and air conditioning application is different. For instance if you are working on ice machines essentially, you wouldn't have the need for the tools that are being used by a service technician that is servicing window units. In this chapter, I've compiled a general list of basic tools that should be in all technician's tool boxes. The specialty tools are not listed.

MULTIMETER (FIG. 35-1)

A good meter is very important, it is the tool that enables you to see the invisible, electricity, in different forms. Much of your diagnosis will depend upon the accuracy of this tool. These meters can range in price from $15.00 to $200.00. The meters have different scales. If you need high voltage scales, you will need a better than average meter. The meter should also have a case, and the tool should be treated with care. The more it bounces around in the back of a truck unprotected, the less its accuracy and the shorter its life span.

CLAMP-ON AMMETER (FIG. 35-2)

The ammeter is very important. This instrument enables you to measure the actual amperage a device is using. Each manufacturer

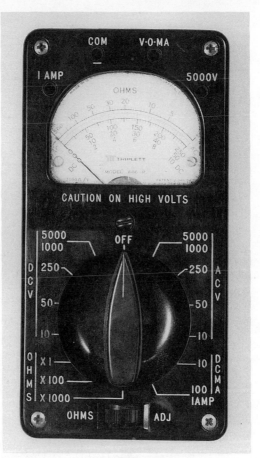

Fig. 35-1. Multimeter.

has a safety limit attached to his product that tells the service technician the limits of amperage to be used in the specific appliance or device. Without an ammeter, it is impossible to know if a unit is operating properly.

MANIFOLD AND GAUGES

This is another tool that is a must. You cannot service anything without knowing its operating pressures. The two gauges that come mounted on the manifold are standard. The only thing I want to mention is the need for long enough hoses. The three hoses that are standard with the manifold assembly are usually about 36 inches long. These at times make the hook up a little difficult. For this

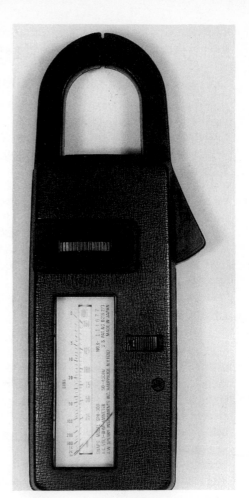

Fig. 35-2. Clamp-on ammeter.

reason, at the time of purchase consider getting your manifold equipped with hoses about five or six feet long. Hoses can be purchased at any length you want. Of course the increase in length causes a proportional increase in price. In some cases this is a fact not to be considered. For instance if you are working on the refrigeration plant that is mounted on the front of a trailer being pulled by a 10-wheeler, it is easier having a large drum of refrigerant on the ground with a long hose attached to it. There are also cases when refrigerant drums are too heavy to carry from the truck. With the use of long hoses, the drum can be left on the truck, and the refrigerant can be delivered any distance away. I also carry a spare

311

manifold and hose seals. The approximate cost for this tool is $40.00 and up, depending upon the length of hoses.

OXYGEN-ACETYLENE SOLDERING OUTFIT

This tool is another one that is indispensable. If you are a service technician or installation technician, you must have a set of torches. The outfit will cost about $150.00. The outfit usually includes oxygen gauge with regulator, acetylene gauge with regulator, welding glasses for eye protection, striker used for ignition, cutting handle and tip, #2 and #5 tips with handle, and siamese hoses. The actual steel bottles that attach to your regulators are rented from the supply house. A sizable deposit is placed on the bottles during the initial purchase. The contents of the bottle are relatively inexpensive whether is it oxygen, acetylene, or nitrogen. The deposit amount depends on your area. The sum of $50.00 would be about the average.

LEAK DETECTOR

The halide detector is the old standby. The price is a moderate one at about $35.00. Many technicians are turning away from the halide detector and are using electronic leak detectors. They have their good points and bad points. They are exceptionally sensitive and excellent in finding minute refrigerant leaks. The cost of the electronic leak detectors have varied as does most electronic products. They were $130.00 and then many were introduced onto the market which brought some of the prices down. With this piece of equipment, you are dealing with electronic circuitry which can present problems if an inexpensive model is purchased.

ELECTRIC DRILL

The ⅜-inch drill has an initial cost that might be slightly higher than a ¼-inch drill . . . the difference in price is well worth having the ⅜-inch drill. The cost of this tool will average about $35.00. Remember that the area you are in will determine the actual pricing of all tools and equipment. The prices I quote are only given as a general overall average.

ELECTRIC SCREWDRIVER

This tool has expedited assembly and disassembly of the average air conditioning unit. These tools are made to operate as a portable having batteries that are chargeable, or electricity with the use of a service cord. The batteries will not last all day if you are doing installation work. The portable unit works well for service.

SOLDERING (IRON-GUN)

This tool is invaluable when doing a lot of control work. If you do more general service work, this tool will not be used too often.

ADJUSTABLE WRENCHES (FIG. 35-3)

These wrenches come in a variety of sizes. Most technicians use the eight and 10-inch sizes mostly. Of course, the longer you

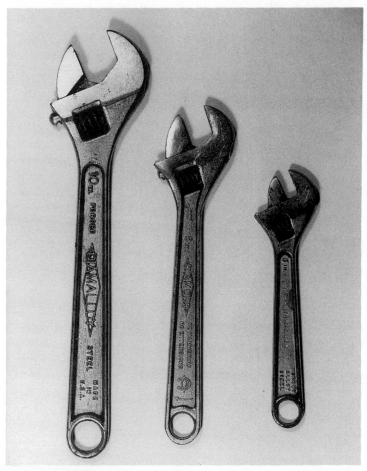

Fig. 35-3. Adjustable wrenches.

313

are in servicing, you will eventually want to purchase a set of them up to 16 inches.

NUT DRIVERS (FIG. 35-4)

A set of nut drivers costs about $15.00 and lasts for a long time, unless you are like some technicians that leave a tool on every job. I also say at this time that there are a lot of inexpensive tools on the market that are manufactured to look like professional tools; however, under normal use, they fail terribly. It is prudent to purchase a better quality tool in the beginning and save yourself a lot of time and aggravation. Try and stay with the name brand tool manufacturers. Nut drivers will be used constantly; buy good ones.

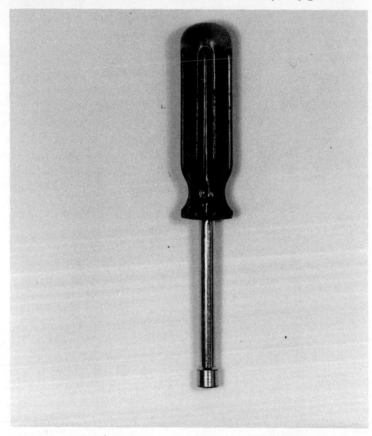

Fig. 35-4. Nut driver.

STANDARD SCREWDRIVERS

A few good standard screwdrivers are a must. The shaft length and blade width should vary giving you a selection. Again, buy quality, for these tools are in constant use. We all know that these screwdrivers sometimes double for chisels, most of us are guilty of this action, for that reason get good drivers.

PHILLIPS SCREWDRIVERS

The same advice applies for these screwdrivers as the standard ones, buy quality ones so that the tips will sustain a long life. When inexpensive Phillips screwdrivers are used, their ends deteriorate very quickly due to the fact the screw head is made of a harder metal than the tool. This is not the way it should be.

ALLEN WRENCHES (FIG. 35-5)

These wrenches are the ones used to remove the recessed hex head set screws. They are made of case hardened steel and will sustain extreme pressures without breaking. The wrenches come in a variety of sizes. Some have long shanks for use on window units when the squirrel cage fan must be removed. The conventional L-shaped wrench is made in different lengths. A small tool packet has the popular sizes hinged with a common pivot pin and encased in a metal enclosure that looks like a fat pocket knife. This tool set is convenient to some technicians. Another wrench set has a common handle with long lengths of hex-stock. Prices for this tool are not predictable as the screwdrivers above can't be priced. There are too many determining factors.

THERMOMETERS

Our business is based on controlling temperature, for this reason a good thermometer is a necessity. The problem is which one do you want to use. The least expensive is the alcohol-filled thermometer. It is fairly accurate and costs about $4.00. The next thermometer up the ladder is the mercury-filled thermometer, that has a cost of about $8.00. These units are fairly accurate, and the only objection to them is they are very fragile and are slow in producing their final readings. The dial thermometer has become very popular. The name brands are fairly accurate and hold their adjustment. The dial face is easy to read and has a needle that deflects

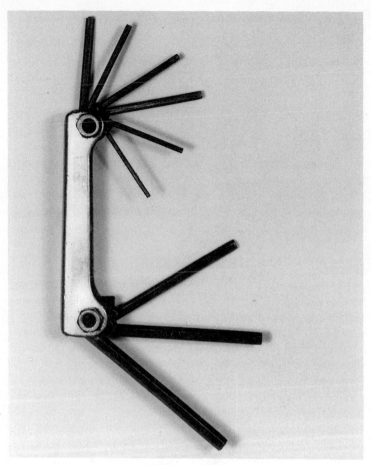

Fig. 35-5. Allen wrenches.

with the temperature changes. It is not too fast responding to temperature changes. The unit has a calibration adjustment on it that makes accuracy constant, if the technician takes the time to check his thermometer in ice water. This doesn't guarantee that accuracy will be maintained in all temperature ranges; however, it will detect a defective thermometer that should be retired.

Again, electronics has made its appearance in the newest thermometers. For about $80.00 to $150.00, you can own a fast-acting, accurate thermometer. Some of these units are made to accommodate two or more probes which allow for the reading of more than one temperature at a time. This is a valuable asset when working in refrigeration and some air conditioning.

Recorders can also be purchased for about $250.00 where you can accumulate the constant temperature of a unit over a long span of time. Units are designed to measure from 24 hours to seven days. Of course this piece of test equipment will not be found in the average technicians tool box, I'm only explaining its availability.

PLIERS (FIG. 35-6)

Two types of pliers to have are a regular pair of pliers and a pair of slip-joint pliers (sometimes called water-pump pliers). These tools are not expensive and can be purchased for about eight dollars each. Electricians pliers are more expensive due to their hardened cutting edge. The latter is a good tool and is essential if you do a lot of wiring. Along with this tool, a pair of side cutters or diagonal cutters is good to have. Wire strippers are important and can be bought cheaply.

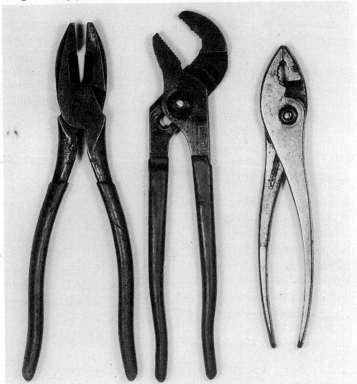

Fig. 35-6. Pliers. Left to right, electrician's, water-pump, slip-joint.

HAMMER

Yes, a hammer, you don't use your adjustable wrenches for a hammer. You'll need the hammer for many reasons. I feel the claw hammer is most suitable for our industry. Many prefer the ball-peen hammer.

SOCKET WRENCHES

A good set of sockets and a ratchet will last for many years. A ¼-inch and ½-inch drives should be acquired. Both sets will give many years of service.

TUBING CUTTERS (FIG. 35-7)

Two tubing cutters are desirable. Get both the standard size the and the midget cutter that opens enough to cut ⅝ inch tubing close to the wall. The cutting arc of the midget cutter makes cutting in close spaces easy and possible. The standard cutter might require too much of a radius in certain instances.

FLARING TOOL (FIG. 35-8)

This two-piece tool is another one that will be used frequently. There are many different designs of flaring blocks, so it is up to the individual to make the selection. This tool will cost about $25.00.

SWAGING TOOL

Swaging is another method of making a joint in copper. If a coupling is not available, or you just want one joint instead of the two that is necessary with the coupling, the swaged joint is used. This method also saves on materials without having to use the coupling. The principle of swaging is to stretch one piece of copper tubing to a point that the o.d. (outside diameter) will slip into the i.d. (inside diameter) of the stretched piece. This is accomplished with the aid of swages. They are made for ¼-inch and larger size copper tubing. A hammer is used for the driving force in swaging the copper.

INSPECTION MIRROR

This mirror is similar to the one used in the dentist's office, it enables the user to see behind objects such as piping. This tool is a must when soldering in blind places behind pipes or fittings. The mirror allows you to see the finished joint and correct any leaks,

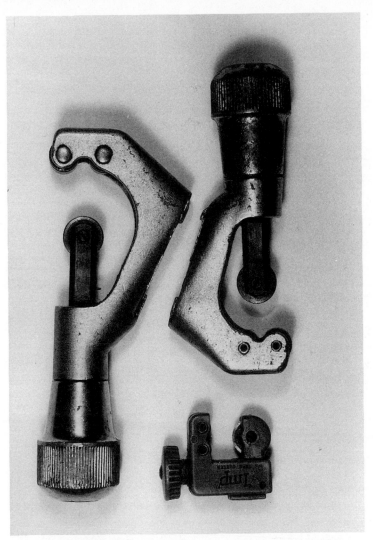

Fig. 35-7. Tubing cutters.

if necessary. It is much easier to make a good joint the first time than to come back on a service call. The mirror is also an aid to seeing fittings where oil leaks are developing.

SERVICE-VALVE WRENCH (FIG. 35-9)

This wrench is a must also. Many have used an adjustable or other type of wrench to open this type of valve and totally destroyed

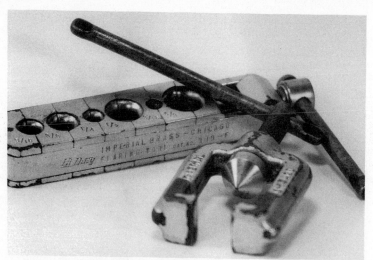

Fig. 35-8. Flaring tool.

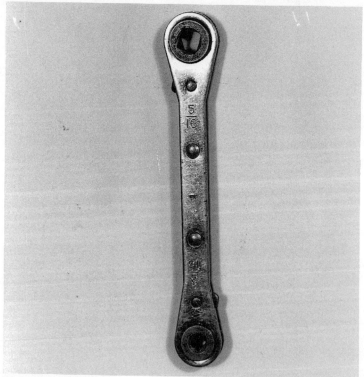

Fig. 35-9. Service valve wrench.

the valve stem that has a square machined in it to accommodate the service-valve wrench. Eventually, a pair of pliers are used for the valve. The wrench comes in several styles and is not expensive.

CRIMPING TOOL (FIG. 35-10)

When the sealed system of an air-conditioning system or refrigeration system is serviced, it is sometimes desirable not to leave an access fitting. The access valve is soldered to a long piece of copper tubing. When the refrigerant is placed into the system and the operation has been checked, the system is ready to be sealed as it was in the factory that manufactured the unit. This is accomplished with the use of a crimping tool. The block compress the walls of the copper tubing together very tightly sealing the refrigerant. The valve is cut off the tubing with a tubing cutter. The tubing is then soldered. After the soldering is done, the tool can be removed. There are other versions of the crimping tool; some are designed like vice grip pliers. This tool is not expensive and is worth the investment if you are working on small window units and domestic refrigerators. The cost of installing valves will more than pay for

Fig. 35-10. Crimping tool.

this tool, plus the fact that there will not be a leak potential due to the valve installation. The tool sells for about $18.00.

There are many tools that have to be acquired when you expand your expertise in the air conditioning and refrigeration servicing industry. The few tool listed previously are only some of the specialty ones that you will probably be using. Those tools will only be found in the industry supply houses. Other tools such as hammers and different types of screwdrivers, saws, and files, can be purchased anywhere that tools are sold. As with the meters that I have explained earlier, if you are going to be a service technician and use these tools constantly day after day, buy the better tools. The initial cost is higher but the tools last longer and don't fail at critical times when you will be in the middle of nowhere. The homeowner need not worry about this factor. Of course there are tools on the market that don't even make a good toy for a child due to their short lifespan. For this reason the homeowner should be cautious in buying tools. A tool that looks good and breaks the first time used is neither a tool nor a bargain.

CHAPTER 36

Geothermodynamics

One of the most important things in our industry is to stay abreast and aware of the new developments that take place within it. Computerized control circuits were introduced many years ago, and there are many manufacturers that custom design the special computers to meet the needs of each installation. They should not be feared nor shied away from. They are here and a part of our industry. Remember in the beginning of this book, you were told that you were in a very diversified industry. Your knowledge has to encompass electricity, control wiring, piping, pressure piping, plumbing, mechanical, and now there are a few more items to add to the list.

Of course you will not be expected to rebuild a micro-processor in the field or build micro-chips. You will be required at times to troubleshoot some of these computer-controlled systems. It will depend on what aspect of the industry you intend to service. These units will be found in hospitals, office buildings, industrial factories, to name a few.

The computerized-control panels are shipped for installation with a full set of installation instructions, service manuals and owner's operation manual. They are not that difficult to service once you learn what they are supposed to do and how their circuits are wired.

Many of these systems have been expanded by linking them with the telephone systems. The system has been in a pilot program for several years now. Many of the supermarket chains have had their stores throughout the United States linked to a central computer. Each store then becomes a terminal that transmits the temperature of that store, and readings of temperature from every individual piece of refrigeration equipment can be read. In this way, the central office has total command of all the equipment it owns throughout the country. If a problem is spotted where the temperature of a freezer is beginning to act erratic, the observer can call the local service company in that area and alert them of the problem. In this way, costs of loss of product are kept to a minimum.

This is only one of the developments that has been taking place. Another one that is making progress is geothermodynamics. It began by trying to harness the heat from beneath the earth's surface. In different sections of the country, different applications are being experimented with. Some areas are trying to harness the steam from the earth's depths. This is occurring in portions of the west and southwestern United States. In many southern states, a theory is being worked on at present. Remember that the basics always come into play even during research and development. You have learned that in refrigeration theory, heat of rejection is transported by a medium from an objectionable area to an area where it is rejected from the medium. In air conditioning and refrigeration this medium is a refrigerant being used as a primary transporter and/or water which becomes a secondary refrigerant. The experimentation taking place now is using the earth as a medium to effect heat transfer. Of course the theory is only applicable to the geographical latitudes that do not have a hard freeze. The concept of the theory is to remove the condenser fan motor, blade, and condenser coil from the system. A hole would be dug into the earth about three feet in depth. The exact dimensions of the hole depend on the unit tonnage and are still in the experimental stage. Piping is placed into the hole in a similar arrangement as is the tube-and-tube, water-cooled condenser. Arrangements such as the coil-in-shell are also being tried. The earth is replaced into the hole after the placement of the copper piping. About 12 inches below the surface, different means such as tar paper are being used to keep whatever shallow root system that would be planted above, from being dried out by the heat generated from the discharge line buried below.

The theory of conduction it put into play. The earth at that depth would be about 72 degrees year round. The hot refrigerant gas would give up its heat to the surrounding earth. In turn, the soil would transfer the heat to the soil next to it. If the application proves workable not only will the system be more efficient, it will require less energy to operate. Of course if this system is piped through a heat reclaim unit, more savings will be had by the owner.

It seems that certain manufacturers are already producing condensing units in a different way. The compressor and its needed components are housed in a separate cabinet. The condensing coil and fan motor are housed alone in their own cabinet. This can allow the system to be tailored for different applications. The compressor section can be placed in an enclosed area that would eliminate some of the noise. It can also be removed from the harmful elements that sometimes cause premature wear and deterioration of the compressor and its components.

There are other avenues in which the service technician can become involved. I used to call myself and some of my service technicians, "environmental engineers." It must have sounded good for a few companies took up the name. In our industry we are very concerned with the environment that surrounds people as they work, play, and sleep. Think about ventilating different factories and how proper air changes aid in better health to those working within that environment. For this reason, I'm going to include just an idea of solar power in the next chapter. Many companies who basically are in the air conditioning, heating, refrigeration, and ventilating business have expanded into the solar energy market.

CHAPTER 37

Solar Energy

The service technician has been aware of solar energy for many years. Ask those of us that have worked on a rooftop unit and placed an adjustable wrench on the unit, then reached for it in a few minutes. We dropped it like a hot potato, for it was. How about those times when a wrench was placed next to your knee and the roof was so hot that the asphalt turned to liquid again and the wrench sank. Yes, many of us are very aware of solar energy. Only a few years ago it appeared on the market as if it were something new.

Although some are interested in creating electrical energy with the use of solar collectors, those in our industry have developed systems to create hot water for the domestic water supply of a home. It can also be designed into the heating system. In certain areas, hot air heat collectors were experimented with. Of course many of these systems are only proto-type systems; however, many systems have been sold to the consumer, and many more will be sold. For this reason I believe that you should have a basic working knowledge as to how the solar collectors operate.

In Chapter 33, heat reclaim was discussed. The solar collector is very similar, except it uses the heat from the sun to supply the needed energy instead of the heat of compression and discarded heat from a controlled environment. It is my opinion that these units will become cost effective and inexpensive to install, thus bringing many units into the average household. Naturally, the more sun that is available the faster the area will accept solar powered devices.

COLLECTOR

This is the unit that appears on the roof or in the yard where the sun's rays can strike the collector at the proper angle. With different types of surfaces being used including prisms, the angle of entrance varies. The location should afford the most time exposure to the sun.

The construction of the collectors varies from plain plastic black tubing stretched across a roof, to complex hi-tech prism chambers. The main purpose of this unit is to absorb the heat from the sun's rays. This heat is then transferred to a medium that transports the heat to a storage area. In most cases, water is used for the domestic hot water consumed in a structure.

CIRCULATING PUMP

A small circulating pump is used to circulate water to the collector and back to the storage area continually. This procedure continues to elevate the heat of the water until the desired temperature is reached, at which time the pump will stop operating. It is easy to see how the hot water heater of a structure can be supplemented with the help of solar energy. The solar heat must be present in order for the system to be operational.

HOT WATER SENSING THERMOSTAT

If it is a cloudy day, and there is not enough heat getting through the cloud cover, the water pump will not operate. In some systems there is a solenoid valve that will not open; thus isolating the solar hot water heating system from the conventional hot water heater, under these conditions.

Solar systems can also be used for heating a structure. Again let me say that all of this solar innovation technology is in its infancy, and perhaps this is why I am bringing it into the manual. Research and development is done by all of us. In using solar energy, heat must be constant and there can't be a lack of it like there can be with a water heater when it is recovering from the family showers. Some are trying to answer this problem with an old theory, remember the rule-of-thumb about everything is based on the basics. For this reason, experimentation with the use of thermal-banks is showing some success.

THERMO-BANK

This is not a new idea, and those of you that have been in the industry for a while will remember their use in the low-temperature refrigeration systems. For those that don't remember I'll explain. The thermo-bank is nothing more that a heat exchanger. It is designed similar to a coil-and-shell, water-cooled condenser. The size varies by the size of the system with which it is being used. For example, the heat-bank I'm describing is a square tank about five feet square and five feet tall. Inside the square tank there is a coil of copper. The diameter and length is determined again by the amount of heat needed and how quickly a heat exchange can be effected. The tank itself is filled with a heavy weight oil. There are caps at the top of the tank in order to check and maintain the proper oil level. Oil has a much higher boiling temperature than water, and for this reason, oil is used to absorb the heat from the water when it is leaving the solar collector. This absorption of heat can continue during the times of non-demand on the system. The heat is being deposited into the thermo-bank when there is no need for it. When a demand is made, the heat can be supplied by both the collector and the thermo-bank.

This type of system will work only when there is a hot water heating system. It can also work well if there is a chilled-water system of cooling. Some larger homes have this type of installation. If a structure has a hot air heating system, a water coil can be installed in the discharge air plenum, and with the use of a two-stage thermostat, the solar heat can be used for the first stage and the regular furnace fuel used as a second stage of heat.

This type of system is in the stages that there is no wrong way of application. Whatever works is right, not wrong. For this reason you should familiarize yourself thoroughly with the controls available and what each of their limitations might be. For instance, you would want some type of temperature control to regulate the solar collector. In the event this control fails, you would want some type of pressure relief device that would prevent costly damage. Many of you might want to experiment and try an installation within your own home. Keep it simple and remember to stick to the basics.

HOT AIR SOLAR COLLECTORS

Some have been working with the hot air theory where the air is heated directly by the sun, then circulated into the conditioned space. A return air back to the collector is needed. The system has worked well in many cases. This is something similar to a greenhouse and can be eventually used in northern latitudes. In fact much of the solar experimentation is not being restricted by latitude boundaries.

Prism experiments are being conducted to bend the sun's rays in order to concentrate more heat into a controlled area. It might be something similar to a scout starting a fire with tinder, only using the magnification factor to amplify heat. With this in mind, steam turbines can be driven with solar beams. This would give the world an inexhaustible supply of a non-polluting inexpensive energy. One of you reading this right now might be instrumental in its coming about.

With the development of new refrigerants, absorption-type of air conditioning and refrigeration equipment will become popular again. The use of solar energy could power these systems without the use of ammonia bromides. With the use of proper insulation in the refrigeration equipment, times of no productivity by the solar energy would not affect products contained within the equipment. The insulation would hold temperature. Solar equipment for our industry will probably carry a high initial price tag; however, in seeing the low operating expense and the lack of moving parts that require servicing, solar equipment may cost very little once the installation is made.

Index

333

Other Bestsellers From TAB

Other Bestsellers From TAB

Other Bestsellers From TAB